*Instructor's Guide to Accompany*

# INTRODUCTION TO FIRE PROTECTION

*Robert W. Klinoff*

**Delmar Publishers**
an International Thomson Publishing Company I(T)P®

Albany • Bonn • Boston • Cincinnati • Detroit • London • Madrid
Melbourne • Mexico City • New York • Pacific Grove • Paris • San Francisco
Singapore • Tokyo • Toronto • Washington

**NOTICE TO THE READER**

Publisher does not warrant or guarantee any of the products described herein or perform any independent analysis in connection with any of the product information contained herein. Publisher does not assume, and expressly disclaims, any obligation to obtain and include information other than that provided to it by the manufacturer.

The reader is expressly warned to consider and adopt all safety precautions that might be indicated by the activities described herein and to avoid all potential hazards. By following the instructions contained herein, the reader willingly assumes all risks in connection with such instructions.

The publisher makes no representations or warranties of any kind, including but not limited to, the warranties of fitness for particular purpose or merchantability, nor are any such representations implied with respect to the material set forth herein, and the publisher takes no responsibility with respect to such material. The publisher shall not be liable for any special, consequential or exemplary damages resulting, in whole or in part, from the readers' use of, or reliance upon, this material.

**Delmar Staff**

Publisher: Robert Lynch
Acquisitions Editor: Mark Huth
Developmental Editor: Jeanne Mesick
Production Coordinator: Toni Bolognino
Art and Design Coordinator: Michael Prinzo

COPYRIGHT © 1997
By Delmar Publishers
an International Thomson Publishing Company

The ITP logo is a trademark under license

Printed in the United States of America
For more information, contact:

Delmar Publishers
3 Columbia Circle, Box 15015
Albany, New York 12212-5015

International Thomson Publishing Europe
Berkshire House 168-173
High Holborn
London, WC1V7AA
England

Thomas Nelson Australia
102 Dodds Street
South Melbourne, 3205
Victoria. Australia

Nelson Canada
1120 Birchmount Road
Scarborough, Ontario
Canada M1K 5G4

International Thomson Editores
Campos Eliseos 385, Piso 7
Col Polanco
11560 Mexico D F Mexico

International Thomson Publishing Gmbh
Königswinterer Strasse 418
53227 Bonn
Germany

International Thomson Publishing Asia
221 Henderson Road #05-10
Henderson Building
Singapore 0315

International Thomson Publishing - Japan
Hirakawacho Kyowa Building, 3F
2-2-1 Hirakawacho
Chiyoda-ku, 102 Tokyo
Japan

All rights reserved. No part of this work covered by the copyright hereon may be reproduced or used in any form or by any means—graphic, electronic, or mechanical, including photocopying, recording, taping, or information storage and retrieval systems—without written permission of the publisher.

2 3 4 5 6 7 8 9 10 XXX 02 01 00 99 98

Library of Congress Cataloging-in-Publication Data    96-11630

ISBN #0-8273-7253-1

# Contents

| | | |
|---|---|---|
| Chapter 1 | Fire Technology Education and the Firefighter Selection Process | 1 |
| Chapter 2 | Fire Protection Career Opportunities | 7 |
| Chapter 3 | Public Fire Protection | 11 |
| Chapter 4 | Chemistry and Physics of Fire | 21 |
| Chapter 5 | Public and Private Support Organizations | 27 |
| Chapter 6 | Fire Department Resources | 33 |
| Chapter 7 | Fire Department Administration | 47 |
| Chapter 8 | Support Functions | 57 |
| Chapter 9 | Training | 65 |
| Chapter 10 | Fire Prevention | 73 |
| Chapter 11 | Codes and Ordinances | 83 |
| Chapter 12 | Fire Protection Systems and Equipment | 91 |
| Chapter 13 | Emergency Incident Management | 103 |
| Chapter 14 | Emergency Operations | 111 |
| | Transparency Masters | 121 |
| Chapter 15 | Answers to Student Manual | 165 |

## LESSON PLAN GUIDE

**Instructions:**

This guide is provided to assist the instructor in presenting the material included in the text. The lesson plan guide is, for the most part, an outline of the text.

At the start of each section are the objectives as stated in the text. Key safety points are printed in bold italic type ending with an exclamation point. Audio-visual aids are included for preparation as overhead transparencies and are indicated in the lesson plan by an asterisk. Questions from the end of the chapters in the text and their answers are included at the end of each section.

---

**Online Services**

**Delmar Online**
For the latest information on Delmar Publishers new series of Fire, Rescue and Emergency Response products, point your browser to:
    http://www.firesci.com

---

**Online Services**

**Delmar Online**
To access a wide variety of Delmar products and services on the World Wide Web, point your browser to:
    http://www.delmar.com
    or email: info@delmar.com

**thomson.com**
To access International Thomson Publishing's home site for information on more than 34 publishers and 20,000 products, point your browser to:
    http://www.thomson.com
    or email: findit@kiosk.thomson.com

A service of I(T)P®

# Chapter 1

# Fire Technology Education and the Firefighter Selection Process

## Learning Objectives

*Upon completion of this chapter, you should be able to:*

- Explain the differences between a community college certificate, an associate degree, and a four-year degree in fire technology.
- List the advantages of obtaining a certificate or degree.
- Assess your career potential in the fire service.
- Give examples of work ethics.
- Explain the need for sensitivity to diversity inside and outside of the workplace.
- Describe the different levels and availability of training programs.
- Give examples of different types of personnel development programs.
- List the steps in the selection process and important aspects of each.
- List ways you can prepare for the selection process.
- Explain the purpose and importance of the probationary period.

## INTRODUCTION

The fire technology curriculum is designed to produce a student with a comprehensive background knowledge in the technical aspects of fire suppression and prevention. This course of study can assist you in preparing for jobs other than that of firefighter.

## LESSON PLAN

### Fire Technology Curriculum

    provides technical knowledge
    manipulative at some schools
    jobs other than firefighter also available
        for persons with physical or other limitations

### College Fire Technology Programs

    certificate program
        fewer units (courses)
    Associate Degree
        core curriculum and electives
        variety of courses, depends on local requirements
        *give examples from school catalog
    Bachelor Degree Programs
        Fire Protection Administration
            California State University at Los Angeles
        Fire Protection and Safety Technology
            Oklahoma State University
        Fire Protection Engineering
            University of Maryland
        FEMA Open Learning Fire Service Program
            regional schools

### *Purpose of seeking education*

    professional development
    preparation for promotion
    pay incentives
    completion of probationary period

### Other College Programs

    Public Administration
    Risk Management
    Industrial Hygiene
    Law
    Medicine
    Chemistry

### Career Potential Assessment
- highest moral and ethical character
- not a job to become a hero
    - act as part of a team
- approximately 120 firefighters die each year in the line of duty
    - *see Appendix A: The U.S. Fire Problem
- long hours of preparation
    - must undergo preparation to perform the job
- firefighters have same problems as other people
    - marital strain
    - substance abuse
- Employee Assistance Programs
- Critical Incident Stress Debriefings
    - after incidents
- commitment to physical fitness
- changing role of firefighters
    - expanding responsibilities
    - community programs
    - public relations
- are you ready to enter a burning building when everyone else is leaving?

### Work Ethics and Human Relations
- work and live with other firefighters for duration of shift
- loyalty to coworkers and department
- dedication to duty
- accept hardship without complaint
- able to follow orders

### *no freelancing!*
- ability and willingness to learn
- willing to accept personal responsibility for your actions

### *must have a positive safety attitude!*
- diversity in the workplace
    - women and minorities as coworkers
- Civil Service rules prohibit discrimination
- no sexual or racial harassment allowed
    - cause for termination of employment

### Training Programs
- pre-service
    - explorers

volunteer firefighter
reserve/cadet programs
Emergency Medical Technician qualified program providers
individual fire departments
associations
professional groups
possible preference during hiring process
in-service programs
academy
station, battalion, department level
state
national
first level of training is the academy for most new firefighters
physical and mental performance observed
stage at which most are terminated
large amounts of homework and stress
National Fire Academy
technical courses on national level
model State Fire Marshal training program
*Transparency 1
college-sponsored firefighter training
University of Nevada at Reno Oil Fire School
Texas A&M

## Personnel Development Programs

prepare personnel for promotion
mentoring
firefighters must be generalists
may be specialists as well

## Selection Process

varies with jurisdiction
recruitment of most qualified applicants
possible prerequisites
high school diploma or GED
Firefighter 1 certification
Emergency Medical Technician
Paramedic
application process
advertisement of opening
limiting procedures

*Transparency 2
notification services
pick up application
    ask for supplemental materials if any
completing application
    read job announcement (flyer)
    *Transparency 3
    typed or neatly hand written in ink
    specify areas of experience that qualify you for the position applied for
    attach other materials as needed
written examination
    reading comprehension
    mathematics
    mechanical aptitude
    weight varies
        pass/fail
        percentage value
    preparation texts available
skills test
    tests for job-related mental ability
oral examination/interview
    evaluates education, work experience, and communication skills
    may be 100% of score used for ranking
    make a good first impression
    preparation is important
        practice orals
        writing out answers to common questions
    physical ability\agility
        ability to perform fire fighting-related physical skills
        practice basic types of events
        *see Appendix B in text for example
final interview
    conducted when your name is near the top of the eligible list
    may change ranking on the list
medical examination
    must be in good general physical condition
    free from heart, back, hearing, or vision defects
background check
    searches for prior convictions and problems
probationary period
    last step in selection process
    may contain goals and require testing to complete
        final manipulative/written examination

## Review Questions *(answers appear in bold italics)*

1. The Fire Technology curriculum is aimed at providing the student with what types of skills? ***Primarily technical.***

2. Before enrolling in college classes the student should meet with which college official? ***A college counselor to review program requirements.***

3. What is the basic college degree in fire technology? ***Associate of Arts or Sciences. A Fire Science certificate may be an alternative with lesser requirements.***

4. List two ways to attend fire academies.
   a. ***Through a college program***
   b. ***Through a training officer's association***
   c. ***Through a reserve, cadet or volunteer program***
   d. ***As a newly hired firefighter***

5. What manipulative certification should you have before applying for a position with the fire department? ***Firefighter 1.***

6. Training programs are offered at various levels, such as state and local. List two others. ***National and regional.***

7. List two preservice opportunities for gaining fire fighting experience. ***Cadet/reserve programs and volunteer firefighting.***

8. What is the age requirement for firefighters? ***The minimum for paid firefighters is 18 years of age at time of hire.***

9. List two basic prerequisites in applying for the firefighter exam. ***A valid driver's license and a GED or high school diploma.***

10. What is the first step in the selection process? ***Usually securing and filling out an application.***

11. List two ways to find out when application periods are open for firefighter examinations.
    a. ***Subscribe to a notification service.***
    b. ***Check newspapers.***
    c. ***Place interest cards at personnel/ human resources departments.***

12. List a source of material used to prepare for the written examination. ***Libraries and bookstores.***

13. List two ways to prepare for the physical agility/ability test.
    a. ***Engage in general physical training.***
    b. ***Visit the fire station and use their equipment to prepare.***
    c. ***Become a reserve/cadet or volunteer firefighter, which can provide you with training and practice for manipulating fire-related equipment.***

14. List two ways to prepare for the oral examination.
    a. ***Have firefighters give you practice with a "mock oral."***
    b. ***Consider possible questions and practice your delivery on tape in front of a mirror.***
    c. ***Utilize available video tapes or written material to help you prepare your delivery.***

15. What is the purpose of the probationary period? ***To let the employer assess your performance both mentally and physically.***

# Chapter 2

# Fire Protection Career Opportunities

*Learning Objectives*

*Upon completion of this chapter, you should be able to:*

- Identify fire protection jobs in the public and private fire service.
- List duties and requirements of the position of firefighter trainee and firefighter.
- List duties and requirements of the position of firefighter/paramedic.
- List duties and requirements of fire heavy equipment operator.
- List duties and requirements of firefighter forestry aid.
- Give examples of fire service jobs other than firefighter.

## INTRODUCTION

There are numerous types of jobs involved in the provision of fire and life safety protection to the community. This lesson is designed to introduce you to some of these. As we go through the lesson, think about how you fit into the overall picture and consider some of the opportunities available.

## LESSON PLAN

A fire technology degree opens up a variety of opportunities.

### Public Fire Protection Careers

- first jobs looked at involve firefighting
    - firefighter Trainee
        - entry level position
        - works under close supervision
        - learning process
        - leads to probationary status
    - Firefighter Fire Department
        - entry level position (probationary firefighter)
        - may become eligible for promotion after specified time in rank and successful completion of promotional testing process
    - Firefighter Fire Department Federal
        - civilian position at military bases
        - much like firefighter fire department municipal
    - Firefighter Paramedic
        - advanced medical training
        - increased responsibility
        - sometimes includes pay incentive
    - Fire Heavy Equipment Operator
        - mostly occurs in departments with wildland responsibility
        - operates bulldozers, road graders, and and so on
    - Safety Section Retirement
        - based on the fact that some jobs are very stressful
        - primarily fire and police
        - higher level of benefits
            - higher percentage at retirement
    - Firefighter (Forestry Aid) Wildland GS 3
        - position with the U.S. Forest Service
        - summer job for many college students
        - provides firefighting training
- Non-firefighting jobs
    - may be titled differently in different departments
    - Fire Prevention Specialist

may be civilian or safety section
  requires knowledge of codes and ordinances
Fire Hazardous Materials Program Specialist
  requires chemistry or related degree
Fire Department Training Specialist
  may be civilian or safety section
  requires education in preparation of training
    materials
    programs
Public Fire Safety/Education Specialist
  may be civilian or safety section
  ability to teach to all age groups
  plan and develop programs for target audience
Dispatcher
  may be civilian or safety section
  firefighter position in some departments
  receives calls and dispatches equipment
    emergency and non-emergency
      may be trained in Emergency Medical Dispatch

## Private Fire Protection Careers

Firefighter
  at industrial facilities
    loss prevention and control
Insurance companies
  loss prevention
  investigation
  claims adjustment
Industry
  loss prevention
  safety
Fire Protection Systems Engineer
  design fire protection systems
Fire Protection System Maintenance Specialist
  service portable extinguishers
  service fixed protection systems
Inventors
  constant flow of new demands on the fire service
  identify need for new technology or procedures
    to do the job more safely or more efficiently
  develop materials and methods to satisfy needs

## Review Questions  (answers appear in bold italics)

1. What is the difference between the firefighter trainee and firefighter positions? *The firefighter trainee is a position that has been created to help a person develop the skills to gain employment as a firefighter.*

2. What is the meaning of the term "under supervision" as it relates to being a firefighter? *When firefighters are "under supervision," they are closely monitored for their safety and training as well as to assess their potential.*

3. What additional requirements are placed on a firefighter paramedic over that of a regular firefighter? *The training required to become a paramedic. In many areas this training takes approximately six months to complete. The position also requires advanced study and learning skills.*

4. What particular skills are required of a fire heavy equipment operator? *The ability to operate heavy equipment under fire fighting conditions.*

5. What are the differences between a safety section and a general retirement? *Safety section retirements promise a higher rate of return upon retirement. They are usually reserved for law enforcement and fire personnel.*

6. As a Firefighter (Forestry Aid) GS 3 can you expect to have a full-time job? *Not initially. It usually takes several years to be offered a position with a guaranteed duration.*

7. What are the benefits of having a summer job as a firefighter (forestry aid)? *You are free to attend school in the winter. You may also get to see much of the country and earn overtime pay during a busy summer. You will also gain fire fighting experience.*

8. What jobs are available in the fire service that do not require actual fire fighting to be performed? *Dispatcher, paramedic (in some departments), fire prevention inspector, and arson investigator.*

9. What are some of the private sector jobs that you could have that would help you to prepare for the position of firefighter fire department? *Fire inspector, industrial fire brigade, emergency medical technician, or paramedic.*

10. How would you go about identifying the need for a new device or procedure for the fire service? *Observing how jobs are currently performed. Many times firefighters have already adapted some tool to perform the job more quickly or easily, but it takes someone to manufacture the tool and bring it to wider attention.*

# Public Fire Protection

## Learning Objectives

*Upon completion of this chapter, you should be able to:*

- Identify the origins of modern fire protection.
- Describe the evolution of fire protection.
- List the causes of the demise of the volunteer fire companies in the major cities.
- Identify the U.S. fire problem as it currently stands.
- List the general responsibilities of the modern fire service.
- Describe the evolution of modern fire fighting equipment.
- Describe the evolution of protective clothing and equipment.
- Describe how major fire losses have effected the modern fire service.
- List the reasons for fire defense planning.

## INTRODUCTION

Organized fire protection has gone through an evolutionary process ever since it started. By studying this process, we can better understand how we got to where we are today and where we are going in the future. This lesson will also take a look at the evolution of the equipment in use today.

## LESSON PLAN

### Evolution of Fire Protection

- first recognized fire fighting force was the Vigiles in the year 6 A.D.
    - equipped with buckets and axes
    - patrolled the streets
    - reminded citizens to be careful with fire
    - fought fires as necessary
    - those who caused fires were punished
- Jamestown settlement
    - structures made of combustible materials
    - chimneys of brush and sticks or wood covered with mud or clay
    - closeness of structures and combustible construction aided fire spread
    - fire protection consisted of pulling down structures to create a fire break
    - bucket brigades were used to transfer water to the fire
- Peter Stuyvesant, governor of New Amsterdam (New York)
    - created building code
        - prohibited wood chimneys
    - appointed fire wardens ("Rattle Watch")
        - levied fines for violations
    - use of curfews
        - fires to be covered or extinguished between hours of 9 P.M. and 4:30 A.M.
- Great Fire of London, 1666
    - burned for 5 days
- fire insurance companies were created
    - companies protected the properties they insured with their own fire fighting forces
    - fire marks placed on buildings
    - properties they did not protect were allowed to burn
    - *Transparency 4
- Boston established first publicly funded, paid fire department in 1679
    - mutual fire societies organized
        - volunteer organizations that assisted the fire department
- Ben Franklin organized the Union Volunteer Fire Company in 1739
    - considered to be the concept used by volunteer fire companies still in existence
- water systems consisted of hollow logs with wooden plugs
    - plug was pulled to access water, giving rise to the term "fire plug"

volunteer fire companies grew in popularity
- equipment became very ornate
- advent of uniforms

companies would fight to see who could get "first water"

protective equipment was not yet invented
- firefighters suffered burns and other injuries

competition among companies helped to lead to their demise in the large cities
- Cincinnati established first fully paid fire department in 1853

expanding role of modern fire departments
- fire prevention
- public safety education
- medical aid
- rescue services
- hazardous materials response

fire service traditions
- have to prove yourself on the fire ground
- engines painted red
- entry at the bottom of the organization

firefighters need to be adaptable
- change is inevitable

change from suppression orientation to medical aid
- public's rising expectations of service

change in fire service education system
- rise of professionalism
- national and state standards for qualifications
- certification and degrees
- possible future name change for firefighter
    - public safety specialist
    - emergency services technician

## Equipment

evolved due to demand for greater capability

siphona, first known fire pump

*Transparency 5
- pump design used in hand pumpers
- pump handles called brakes
- many designs were used
    - two and four cylinder
    - end stroke and side stroke
    - double deck
- first designs discharged water through a gooseneck
- hose developed to aid in water application

away from pumper
to seat of fire
made interior attack possible
original hand pumpers required tub to be filled by hand
attached drafting hose let them draw water from static source
attached suction called "squirrel tail"
hose companies developed
could lay hose from source to pumper
could lay hose from pumper to fire
hand pumpers were pulled to fire scene by personnel
many were large and heavy
crew injured in accidents on way to scene
invention of the steamer
originally pulled to scene by hand
harnessed to horses because of its weight
required only a three-man crew to operate it
could pump for hours
helped cities to move to paid departments due to reduced need for manpower to operate
dalmatian dogs were kept to calm the horses
introduction of ladder companies
ladders carried on a wagon
to gain access to upper floors because buildings were built higher
aerial apparatus developed in 1870
extended by hand cranks
next came spring-assisted, compressed air, and hydraulic
introduction of chemical wagon late 1800s
carried two tanks—one of soda and water and the other of acid
when mixed, created $CO_2$, which created pressure and expelled water through hoses
limited by amount of water in its tank
useful on small fires
if hose was plugged or kinked an explosion could occur
attachment of apparatus to internal combustion powered tractor
led to replacement of horses used to pull apparatus
finally replaced hand, steam, and chemical pumpers

## Fire Stations

developed as departments hired full-time firefighters
replaced sheds for housing equipment
used to house crews 24 hours a day

### Personal Protective Equipment

    developed out of necessity
- first volunteers had no protection
- uniforms developed to identify membership in a company
  - worn in parades and at social events
  - large letters or numbers on front of shirt
  - paid departments adopted uniforms in the early 1800s
  - badges developed to indicate rank and department
- helmet developed to give protection from falling debris
  - long rear brim to keep debris from going down back of firefighter's neck
- turnout gear developed to give protection from heat and water
  - lined to protect from heat
  - boots protected the feet
  - could be worn over regular clothing
- today SCBA (self-contained breathing apparatus) in common usage
  - protects respiratory system
  - offers reduced exposure to toxic atmospheres and low oxygen concentrations

### Fire Losses

- conflagrations have hit almost all major cities in the past
- Baltimore, MD, 1904
  - out-of-town companies unable to connect to the hydrant system
  - led to development of the National Standard Thread
- San Francisco, CA, 1906
  - major earthquake
  - resulting fire was even more destructive
  - 450 killed, 4.7 square miles of the city burned
- Chelsea, MA, 1908
  - major fire destroyed one-half of the city
  - happened again in 1973
- common denominators
  - combustible construction
  - narrow streets
  - unprotected vertical shafts

### The U.S. Fire Problem Today

- combustible roofs contribute to conflagration size fires
  - especially in California
- Santa Ana wind condition
  - high pressure over Great Basin, low pressure over Pacific Ocean

hot, dry winds blow over Southern California

happens in the fall after fuels have had all summer to dry out

fire moves too fast to get in front of

combustible construction just adds to the fuel available

Berkeley, CA, 1923 and again in 1991

    consumed an average of one home every 13 seconds

    deaths: 23 civilians, 1 policeman, and 1 firefighter

*Note:* Most of the civilians and the policeman were killed when they were trapped on a narrow road and the fire rolled over them. They were trapped because a car fell down the hillside when its garage burned out underneath it. The car blocked the road and the vehicles trying to escape were blocked. They were unable to turn around and go in the other direction to escape.

Santa Barbara, CA, 1990

    Paint fire

    545 structures

    wind-whipped fire jumped eight-lane freeway, burning downhill

Fire Siege of 1993, October 26 through November 7

    Southern California

    1,152 structures

    200,000 acres

    burned until winds subsided and/or until the fire reached the ocean

    similar fire experience occurs as other areas build homes in the urban interface/intermix

*Note:* As of 1996, several insurance companies have started to refuse to write fire policies in certain areas. The criteria used are presence of pine and eucalyptus trees, mass brush, narrow access roads, and shake roofs. At this time, it is uncertain whether this will become a trend in the industry.

arson is a cause of many high-loss fires

    firefighter shot in the neck during riots in Los Angeles, CA in 1992

    862 structure fires

    3 deaths

    Detroit, MI, "Devil Night," 1994

        88 structure fires

        25 vehicles

the United States has one of the highest fire death rates per capita in the industrialized world.

    approximately 5,700 die in fires each year

    29,000 civilians injured

    fire usually kills more Americans than all other natural disasters combined

    third leading cause of accidental death in the home

    approximately 70% of deaths occur in residential fires

    more than 2 million fires reported a year

    direct property loss of $8.5 billion

senior citizens at the highest risk of death by fire, double the younger population

people under the age of 19 account for 25% of fire deaths

children under five have double the average risk of being killed in a fire

25% of fires that kill children are caused by children playing with fire

careless smoking is the leading cause of residential fire deaths

in commercial properties, arson is the major cause of deaths, injuries, and fire loss

cooking is the leading cause of apartment fires and second leading cause of single family residential fires

top five areas of fire origin in residences:
- kitchen, 29%
- bedroom, 13%
- living room/den, 8%
- chimney, 8%
- laundry area, 4%

working smoke detector doubles a person's chances of surviving a fire

approximately 90% of U.S. homes have at least one smoke detector

nearly half the residential fires and three-fifths of residential fatalities occur in homes with no smoke detectors

burned businesses often do not recover after a fire, and jobs and tax revenue are lost

lightning is the cause of many large acreage fires

in 1990, 5½ million acres burned

$270 million spent on suppression

the other common cause is humans

60% of the fire starts due to
- unattended campfires
- burning trash
- smoking
- faulty equipment
- driving vehicles with catalytic converters in dry grass

26% caused by arson

## Purpose and Scope of Fire Agencies

original purpose: "to save lives and property from fire"

created large losses from water damage

led to introduction of salvage work

introduction of increased rescue responsibilities
- underwater dive team (SCUBA)
- urban search and rescue (USAR)
- technical rescue
- swift water rescue
- cliff rescue

    auto extrication
    advanced cardiac life support (ACLS)
  fire prevention
    expanded role
    before, during, and after construction and occupancy of buildings
    public education
      safety
        earthquake
        tornado and hurricane
        swimming pool
hazardous materials
  response
  control and reporting
arson investigation
  prosecution
  cost recovery

### Fire Defense Planning

  to provide an integrated fire defense system at the least cost
    what is the acceptable level of loss due to fire
      deaths
      injuries
      property
  *explain the difference between: goals, objectives, policies, and procedures
    goal: a general statement of a desired result
    objective: statement of measurable results to be achieved with resources available
    policy: broad statement used to guide decision making and actions
    procedure: a specific statement of how work is to be performed
  four goals of fire protection
    to prevent fire from starting
    to prevent loss of life when fire does start
    to confine fire to its place of origin
    to extinguish fire once it does start
  first need to determine where the problem lies
    statistics gathered on fire loss
      type of loss
      type of occupancy
      time of day
      ignition source
      item first ignited
      direct cause of loss
look for trends and major factors

used to determine objectives

goals must be politically and financially attainable

*use an example

to achieve the objectives, resources must be provided

    tools and equipment

    personnel

    facilities

cost determination to provide desired level of service

    chief and staff prepare budget

    reviewed by budget analysts

    seek funding from legislative body

once budget is approved

    funds applied to achieve results

    results evaluated and plan changed to address issues

### The Future of Fire Protection

more towards fire prevention

    influence of outside factors

        insurance companies

        tax structure

        advanced technology

        construction methods

        public demands

        federal and state regulations

        OSHA (Occupational Safety and Health Administration)

The one thing for certain is that the fire department of the future is going to be different than it is today.

## Review Questions (answers appear in bold italics)

1. What was the first organized force of firefighters called? ***The Vigiles.***

2. Why was organized fire fighting developed? ***As people moved into towns and lived closer together, the need for organized fire fighting evolved to prevent conflagration.***

3. List several of the measures developed to prevent fires in the early United States.
   a. ***The threat of corporal punishment***
   b. ***Chimney construction requirements and inspections enforced by fire wardens***
   c. ***The use of curfews which required all fires to be extinguished or covered for certain hours of the night***

4. List reasons for joining a volunteer fire company.
   a. ***A desire to help one's community***
   b. ***A sense of pride and accomplishment***
   c. ***The community adoration of those who fought fires***
   d. ***The excitement of fighting fires experienced by the volunteers***

5. What factors led to the demise of the volunteer fire companies in the large cities?
   a. ***The fierce competition among fire companies***
   b. ***The invention of the steamer, which reduced the manpower needed to operate***

  *the fire pumping equipment, i.e., hand pumpers*
 c. *The desire of the community to have 24-hour-a-day protection*

6. Trace the development of fire department pumping apparatus
*The first pumping apparatus was the "siphona"; this principle was applied in the hand pumpers. The next leap forward was the steam-operated pumper, commonly called a "steamer." With the invention of gasoline and diesel engines, the modern pumper was born. Along the way the chemical engine was tried, but it provided limited effectiveness because of its small tank.*

7. Which groups are at the highest risk of being killed in fires? *The very young and the very old.*

8. Where do most fatal fires occur? *Single-family dwellings.*

9. What is the leading cause of residential fires? *Smoking, followed by arson and heating equipment.*

10. What factors have caused the public to expect more services from the fire department? *The increased demand for services is partially caused by television shows, such as "Emergency," that lead people to have rising expectations. Another factor is the transformation of the population from one that was once agricultural in nature and somewhat self-sufficient to one that is urban and depends on services provided by others.*

11. What services are provided by the fire department in your area? *The following are just some of the possibilities: fire fighting, rescue, public education, inspection, emergency medical service, hazardous materials response, hazard reduction, enforcement, arson investigation, and public service, i.e., vehicle lockouts and installation of smoke detectors.*

12. What developments have taken place in the area of personal protective equipment for firefighters? *The personal protective equipment for firefighters has evolved and become specialized into the main types of incidents requiring protection. There is specialized clothing and breathing apparatus used when combating structure fires, equipment for wildland fires, hazardous materials suits, and equipment for emergency medical service.*

13. What are the common denominators in most fires of conflagration proportion?
 a. *Weather*
 b. *Combustible exterior construction*
 c. *Dry ground fuels*
 d. *Spacing of structures*

14. What are the four goals of fire protection?
 1. *To prevent fires from starting.*
 2. *To prevent loss of life in case fire does start.*
 3. *To confine fire to its place of origin.*
 4. *To extinguish fire once it does start.*

15. What is the difference between a policy and a procedure?
*A policy is a statement of general guidelines. A procedure is used to describe how employees will perform important and recurrent functions.*

# Chapter 4

# Chemistry and Physics of Fire

## Learning Objectives

*Upon completion of this chapter, you should be able to:*

- Define the difference between the fire triangle and the fire tetrahedron.
- Describe what constitutes an oxidizer.
- Describe what constitutes a fuel.
- Illustrate the states of matter.
- Explain the process of pyrolysis.
- Describe the properties affecting solid fuels.
- Describe the properties affecting liquid fuels.
- Describe the properties affecting gas fuels.
- Differentiate heat and temperature.
- Illustrate the four methods of heat transfer.
- Illustrate the four classifications of fire.
- Describe the three phases of fire.

## INTRODUCTION

To effectively prevent fires from starting, and to control them once they do start, it is necessary to understand how fires burn, how they start, and what can be done to contain them. This lesson is designed to inform you about the chemical and physical properties of fires. Once we have a basic understanding of all of the factors involved, informed decisions can be made.

## LESSON PLAN

### Fire Defined

> fire: rapid, self-sustaining oxidation process accompanied by the evolution of heat and light in varying intensities
>
> combustion: a chemical reaction that releases energy as heat and, usually, light
>
> rusting of iron is slow oxidation

### Fire Triangle

> three sides
> > original
> > > *Transparency 6
> > > > fuel, heat, and air
> >
> > new
> > > *Transparency 7
> >
> > fuel, energy, and oxidizer
>
> fire tetrahedron
> > *Transparency 8
>
> fuel, energy, oxidizer, and chemical chain reaction

### Chemistry of Fire

> oxidizer
> > oxygen most common
> >
> > occurs as 21% of air
> >
> > increasing amount of oxidizer may increase intensity of fire
> >
> > other oxidizers
> > > fluorine and chlorine
>
> fuel
> > defined as: anything that will burn
> >
> > most common fuels contain carbon and hydrogen
> >
> > complete combustion yields $H_2O$ and $CO_2$
> >
> > most combustion incomplete due to several factors
> >
> > fuel size, arrangement, contaminants, and lack of sufficient oxidizer yields smoke, CO, and other fire gases

### Physics of Fire

- fuel
  - occurs in the three states of matter
  - *Transparency 9
  - solid, liquid, and gas
  - state is often temperature dependent
  - takes both fuel and oxidizer in gaseous state to combine
  - fuel is vaporized by input heat
  - process is called pyrolysis
  - as heat is added, molecule broken into smaller components
    - vaporize and recombine with oxidizer
  - when fuel is hot enough to self-sustain combustion it is at its ignition temperature
- solid fuels
  - factor affecting rate of pyrolysis
  - size, arrangement, continuity, and moisture content
- flame spread
  - Steiner Tunnel Test
    - 25-foot by 20-inch specimen face down on ceiling of tunnel
    - flame applied to one end and drawn by fan for $4\frac{1}{2}$ feet
    - spread measured to point 25 feet from starting point
    - asbestos-cement board rated at 0
    - red oak flooring rated at 100
  - test measures flame spread, temperature and smoke density
- liquid fuels
  - flow like water but do not readily separate
  - *Transparency 10
  - specific gravity: the weight of a liquid compared to the weight of an equal volume of water
  - volatility: ease with which a fuel gives off vapors at ambient temperature
  - *Transparency 11
  - vapor pressure: pressure exerted by vapor molecules on the sides of a container
  - boiling point: when the vapor pressure equals atmospheric pressure
  - *Transparency 12
  - vapor density: relative density of a vapor or gas as compared to air
  - flash point: minimum temperature of a liquid at which it gives off vapors sufficient to form an ignitable mixture with air
  - miscibility: ability of a substance to mix with water
- gas/vapor fuels
  - fluid that has neither independent shape nor volume but tends to expand indefinitely
  - *Transparency 13

upper flammable limit: maximum concentration of gas or vapor in air above which it is not possible to ignite the vapors, too rich

lower flammable limit: lower concentration of gas or vapor in air below which it is not possible to ignite the vapors, too lean

flammable range: proportion of gas or vapor in air between the upper and lower flammable limit

classification of gases

flammable and nonflammable

some nonflammable support combustion

oxygen is an example

flammable vapors are not always visible

## Heat and Temperature

heat defined: a form of energy

possible sources of heat:

chemical: breaking down and recombination of molecules

mechanical: friction, friction sparks, and compression

electrical: arc, spark, static electricity, and lightning

nuclear: fission and fusion

heat expressed

British thermal unit (BTU): amount of heat required to raise the temperature of one pound of water one degree Fahrenheit

Calorie: amount of heat required to raise the temperature of one gram of water one degree Celsius

temperature defined: the measure of the hotness or coldness of an object

*Transparency 14

comparison of temperature measurement scales

## Heat Transfer

4 methods

*Transparency 15

conduction: through a medium without visible motion

convection: through a circulating medium

radiation: by wavelengths of energy

direct flame impingement (auto exposure): combination of the others as objects are bathed in flame

Use example of structure fire to illustrate four methods.

## Classification of Fires

*Transparency 16

Class A: ordinary combustibles

Class B: flammable liquids

Class C: energized electrical

Class D: flammable metals

Use example to illustrate that a fire often contains more than one classification.

### Phases of Fire

incipient: oxygen at 21% flame temperature 1800° to 2200° F, smoke and heat released

free burning: heat production increases, spread to other fuels

smoldering: oxygen content below 15%, flame dies out, glowing combustion, area superheated, charged with smoke and fire gases above ignition temperature

*Note:* At this point, if oxygen is introduced, explosive burning can take place resulting in a backdraft

## *Review Questions* (answers appear in bold italics)

1. List the three legs of the old and new fire triangles.
   ***Old: fuel, air, and heat.***
   ***New: fuel, oxidizer, and energy.***
2. Making the triangle a tetrahedron, what is the fourth side? ***Chemical chain reaction.***
3. What is the most commonly occurring oxidizer? ***Oxygen.***
4. The two most common elements in fuels are: ***carbon and hydrogen.***
5. Fuel may occur in any of the three states of matter. What are these states? ***Solid, liquid, and gas.***
6. Define pyrolysis. ***A chemical change brought about by heat.***
7. Define auto and pilot ignition temperatures.
   ***Autoignition temperature is reached when the fuel ignites without an outside ignition source. The pilot ignition temperature is reached when the fuel is ignited by a spark or flame from an outside source.***
8. List the four factors affecting the burning rate of solid fuels. ***Continuity, arrangement, size, and moisture content.***
9. Would a nonsoluble liquid fuel with a specific gravity of 0.8 sink or float when water is added? ***It would float because the specific gravity of water is 1.0.***
10. What happens when a liquid's vapor pressure has reached atmospheric pressure? ***It is at its boiling point.***
11. Will a gas with a vapor density of 1.2 float or sink in air? ***It will sink because the vapor density of air is 1.0.***
12. What affect does the ambient temperature have on the ignitability of a liquid? ***As the ambient temperature rises, the fuel is closer to or exceeds its flash point, making it easier to ignite. The exception would be in a closed container where the vapor concentration exceeded the flammable range.***
13. Which properties make flammable gases such as acetylene so dangerous in fire situations? ***They have wide flammable ranges.***
14. List the four sources of heat energy. ***Chemical, mechanical, electrical, and nuclear.***
15. Water freezes at what temperature Celsius? What temperature Fahrenheit? ***0°C and 32°F.***
16. List the four methods of heat transfer. How do they affect fire fighting operations?
    1. ***Conduction: affects fire fighting operations by heat transfer through solid objects, spreading fire through walls and floors.***
    2. ***Convection: heated gases rise and preheat the upper areas of confinement. Convection can also spread fires through the transport of burning brands onto surrounding fuels.***
    3. ***Radiation: causes spread of fire through the projection of energy onto surrounding fuels, raising them to their ignition temperature.***

**4. *Direct flame impingement (auto exposure): spreads fire from one area to another through direct flame contact.***

17. When a building is burning and it sets the one across the street on fire, the fire is spread by: (two possibilities.) ***Radiation and/or convection.***

18. A fire in a large trash container could contain which classifications of combustibles? ***Classes A, B, and D.***

19. A forest fire, out of control, is burning in which phase of fire? ***The free burning or second phase.***

20. When a fire is burning quietly because it has consumed most of the available oxygen, it is in which phase? ***The smouldering or third phase.***

# Public and Private Support Organizations

## Learning Objectives

*Upon completion of this chapter, you should be able to:*

- Identify types of support organizations.
- Identify the purpose of specific organizations.
- List how these organizations assist the fire service.
- Identify the organization to contact when information regarding a specific subject is required.

## INTRODUCTION

When we have a knowledge of the many public and private support organizations available, it enhances our ability to perform our job of providing fire and life safety to the community. There are numerous organizations that address the many-faceted problems we are confronted with on an everyday basis.

## LESSON PLAN

Author Note: Only selected organizations have been included in the lesson plan. I have determined these to be the ones most commonly encountered by students at this level.

### National and International Organizations

American Red Cross
- provides assistance to victims of disasters
- can also be called upon to assist disaster workers with canteen facilities

Building Officials and Code Administrators (BOCA)
- creator of one of the three model fire codes used in the U.S.

Chemical Transportation Emergency Center (CHEMTREC)
- 24-hour number for gaining information in case of chemical emergency

Chlorine Emergency Plan (CHLOREP)
- for gaining information regarding emergencies involving chlorine
- accessed through CHEMTREC

Factory Mutual Engineering and Research Corporation (FMER/ FM)
- laboratories developed for evaluating fire protection devices and equipment

Fire Marshal's Association of North America (FMANA)
- formed to exchange information on fire prevention and arson

FIRESCOPE
- formed to develop the Incident Command System
- under constant revision as needs arise

Insurance Services Office (ISO)
- formed to gather information to assist in setting fire insurance rates
- publishes *Grading Schedule for Municipal Fire Protection*

International Association of Fire Chiefs (IAFC)
- to further the advancement of the professional fire service

International Fire Code Institute (IFCI)
- subsidiary of the IAFC/Western Fire Chiefs
- creator of one of the three model fire codes used in the U.S.
- used in 29 states

International Computer Hardware Information Exchange for the Fire Service (ICHIEFS)
- on-line service for the fire-related subjects
- established by the IAFC

International Association of Fire Fighters (IAFF)
- largest union organization representing firefighters

International City Management Association (ICMA)
- local government management professionals
- publishes *Managing Fire Services*

International Conference of Building Officials (ICBO)
- develops model building codes
- publishes *Uniform Building Code*

International Fire Service Training Association (IFSTA)
- publishes numerous training manuals for the fire service
- widely used in the fire service across the U.S.

National Fire Protection Association (NFPA)
- members include fire service and industry
- develops standards for fire equipment and systems
- publishes *Fire Protection Handbook*, *National Fire Codes*, and fire service-related textbooks

National Response Center (NRC)
- one-call notification of agencies involved in hazardous materials emergencies

National Wildfire Coordinating Group (NWCG)
- develops training and reference materials for wildland firefighting
- publishes *Fireline Handbook*

Southern Building Code Congress (SBCCI)
- creator of one of the three model fire codes used in the U.S.
- publishes *Southern Building Code*

Underwriters Laboratories Inc. (UL)
- tests, lists, and marks materials tested for safety
- their listing is commonly found on electrical appliances, fire extinguishers, etc.

## Federal Organizations

Department of Transportation (DOT)
- regulates transportation of hazardous materials in the U.S.
- requires placarding of hazardous materials
- publishes *Emergency Response Guidebook* (ERG)

Federal Emergency Management Agency (FEMA)
- created to place federal disaster response coordination and services under one agency
- parent organization of EMI

Emergency Management Institute (EMI)
- provides training for local government personnel in emergency management

National Emergency Training Center (NETC)
- where public organizations meet to seek training in emergency response

National Fire Academy (NFA)
- training organization dedicated to the professional development of fire service and related professionals

provides courses both on and off campus

maintains Learning Resources Center (LRC)

National Firefighting Education System (NFES)

    publishes reference and training manuals on all aspects of wildland fire fighting

U.S. Forest Service (USFS)

    Department of Agriculture (USDA)

    provides fire protection to National Forests and adjacent lands

Bureau of Land Management (BLM)

    Department of the Interior (USDI)

    provides fire protection to BLM and adjacent lands

National Park Service (NPS)

    Department of the Interior (USDI)

    provides fire protection to NPS and adjacent lands

National Interagency Fire Center (NIFC)

    central coordinating center for firefighting resources for wildland fires

    cooperative effort of USDA and USDI

Bureau of Alcohol Tobacco and Firearms (BATF)

    Department of the Treasury

    assists in the investigation of arson and bomb incidents

Occupational Safety and Health Administration (OSHA)

    established to ensure safe working conditions

    starting to take a close look at the fire service, especially after fatal accidents occur

## State Organizations

State Fire Marshal (SFM)

    enforces state fire laws

    promotes fire-related legislation

    develops statewide training programs

State Forestry Department/Department of State Lands

    manages and provides fire protection on state lands

Office of Emergency Services/Civil Defense

    coordinates statewide efforts on large emergencies

    provides equipment to local fire agencies

## Local Organizations

Burn Foundations

    formed to raise money for burn research and treatment and assist burn victims

Police

    provide evacuation and traffic control in emergency situations

    assist in arson detection and prosecution

Building Department
- works with fire department to ensure fire safety in structures

Water Department
- works with fire department to ensure adequate water supply

Planning Department/Zoning Commission
- determines what is built where
- fire department needs input to ensure service is adequate

Street Department
- determines width of streets and weight limitations of bridges
- needs to advise fire department of street/bridge closures
- may be able to provide heavy equipment needed at an incident

Judicial System
- prosecutor's office performs criminal prosecution
- assists with lawsuits

Emergency Medical Service Agency
- oversees local ambulance providers
- facilitates and monitors emergency medical training
- may be involved in determining availability of emergency room beds and directing transportation of victims at mass casualty incident

Community Service Organizations
- possible source of funding for fire prevention materials and fire equipment

## Periodical Publications

sources of the latest information on fire fighting methods, equipment, and issues
*recommended that periodicals be provided as examples
buyer's guide issues are particularly useful

## Review Questions (answers appear in bold italics)

1. What organization would be a source of information about fire equipment? ***Fire Apparatus Manufacturer's Association, National Fire Protection Association, and fire service magazine buyer's guides.***

2. What organization would be a source of information about fire sprinklers? ***Factory Mutual Engineering and Research Corporation, Fire Suppression Systems Association, National Fire Protection Association.***

3. What organization would be a source of information about labor relations? ***International Association of Fire Fighters, state and local Fire Fighters Associations.***

4. What organization would be a source of information about model codes and their adoption? ***International Fire Code Institute, Building Officials and Code Administrators, and Southern Building Code Congress.***

5. When you arrive at a hazardous materials incident you see a placard on a truck and need information as a guide as to how to proceed. What is your information source? ***Department of Transportation Emergency Response Guidebook.***

6. You are at the scene of a hazardous materials spill and need information on the product.

Who do you contact? ***Chemical Transportation Emergency Center (CHEMTREC).***

7. At the same incident the spill is now about to enter a navigable waterway. Who do you contact to alert the proper federal agencies? ***National Response Center and the United States Coast Guard.***

8. The responsible federal agencies have been alerted. Which ones are likely to respond? ***Environmental Protection Agency, United States Coast Guard.***

9. Your department needs training manuals on a variety of fire fighting subjects and you are asked to find a source for them. Who would you contact? ***International Fire Service Training Association and other publishers. These can be easily researched in a catalog from The Firefighter's Bookstore. For wildland firefighting materials, contact the National Interagency Fire Center.***

10. There has been a large-scale natural disaster, other than a fire, in your area. Which federal agency takes responsibility for providing assistance? ***Federal Emergency Management Agency.***

11. Which state level agency is activated to assist? ***It depends on the name of the organization in the state involved. In some it is the Office of Emergency Services.***

12. There is a wildland fire of extreme proportions in the National Forest in the area. Which other federal agencies can be called on to assist? ***The military (both state and federal) and the Federal Emergency Management Agency.***

13. On the preceding fire, if State resources are needed, which agencies are called upon to assist? ***State Department of Forestry/State Lands, National Guard, and Office of Emergency Services.***

14. You are trying to raise money to purchase a rescue tool for your department. What are some of the local service clubs you can contact? ***Lions, Jaycees, Rotary, and others in your area.***

15. Arson has become a growing problem in your jurisdiction. What public and private groups can be activated to form an Arson Task Force? ***Chamber of Commerce, business associations, insurance associations, law enforcement, fire department, and the District Attorney.***

# Chapter 6

# Fire Department Resources

## Learning Objectives

*Upon completion of this chapter, you should be able to:*

- List fire department facilities.
- List advantages of a department having it's own facilities.
- Describe the purpose of each of the fire department facilities.
- Describe the types of fire apparatus and their functions.
- List the types of tools carried on fire apparatus.
- Describe the use of the various tools carried on fire apparatus.
- Describe the different types of personal protective equipment used by firefighters.
- Describe the types and uses of aircraft in fire fighting.

## INTRODUCTION

The modern fire department uses a variety of facilities and equipment to accomplish its mission. This lesson is designed to introduce you to some of the facilities you may encounter in your career. It also introduces you to nomenclature and identifies many of the tools in common use. If you don't know what it's called, it's hard to ask for it.

## LESSON PLAN

### Headquarters

- home of the managerial functions\office staff
- concentrates decision makers in one place
    - facilitates decision making
- may be located at a fire station
    - advantages
        - close to what is happening in operations
        - personnel available to carry out assigned tasks
        - personnel at main station most informed as to what is going on
        - may be beneficial for promotion
        - may get first pick of choice assignments
    - disadvantages
        - noise and distractions
        - disruptive to work routine as assigned by company officer
        - under closer scrutiny of supervisors
- remote location
    - advantages
        - allows for expansion
        - reduced rumors from overheard conversations
        - increased freedom of staff to conceptualize
    - disadvantages
        - less personnel available to carry out routine tasks
        - tends to over-formalize relationship of line and staff

### Automotive Repair Facility

- needed for fleet maintenance
    - mechanics with expertise in fire equipment repair
    - requires hoists for large vehicles

### Training Center

- need not be overly fancy or expensive
    - props constructed from donated items
- drill tower
    - ladder training, aerial and ground ladders
    - high rise training

high angle rescue and rappelling
burn building

**adhere to NFPA Standard 1403** **Live Fire Training in Structures!**
    flammable liquids and tires *should not be used*
    wet straw with a road flare for smoke
    waste wood or pallets for heat
    Safety Officer
    full PPE
    search/interior attack
    fire environment demonstrations
    ventilation
    hazardous materials drills
classrooms
    VCR's, TV's, satellite, cable TV
    better than apparatus rooms
        clean, controlled environment
        share with other agencies
storage rooms
    ladders, hose, etc. dedicated for training purposes
drafting pit, on-site hydrants
    practice and operator testing
    equipment testing
driver training/testing course
    emergency stop and lane change
    away from traffic
hazardous materials props
    plumbing, tanks, and cylinders
confined space rescue props
    tanks
    structure collapse
studio
    duplicating training videos
    creating training videos
    producing closed-circuit TV programs
offices
    for training staff
    duplicating facilities
    instructor support
    library of training materials

### Warehouse/Central Stores

stocks day-to-day needs of department
    toilet paper to turnouts

rebuilding equipment
SCBA maintenance
re-couple hose
SCBA air compressor and oxygen refill station

### Dispatch Center

Computer Aided Dispatch system
- shows address from phone number (enhanced 911 system)
- identifies cross streets
- shows proper response (stations)
- replaces "run cards"
- speeds up dispatch process

### Fire Stations

first were sheds for apparatus storage
have developed into living quarters and offices for department business
new designs incorporate changes for women in the fire service and handicap access
- separate bedrooms
designed so outside blends with neighborhood
should be well maintained
- befitting of professional image
needs large lot for maneuvering apparatus
- secured to deny unauthorized entry
equipment for vehicle maintenance
- air compressor for tires
- vehicle wash rack
hose tower
- for drying hose and salvage covers

### Fire Apparatus

many types required to perform modern fire department functions
built up from cab and chassis
commercial or custom cabs
- designed to meet NFPA Standard 1901 *Pumper Fire Apparatus* specifications
  - requires inside seating for all personnel

### *seat belts* must *be worn, remain seated and belted in until told to exit!*

tour of cab
- large mirrors
- gauges on dash used to monitor equipment status

                air brake pressure
                oil pressure
                fuel, etc.
            manual or automatic transmissions in use
            pump transfer case
                transfers power to pump
            some automatics equipped with a retarder
                acts as brake on drive line
            diesels may have "jake brake"
                acts as compression braking
            both take some of the load off of the wheel braking system
        switches for lights and electronic siren
        lights enclosed in light bar on top of cab
            alley lights for side lighting
            air horns in front bumper
                at auto driver's level
            siren with several tones
        radio system
            multi-channel capability
                administrative and emergency operations channels
                communication link to adjoining jurisdictions
            intercom system
                allows intra-cab conversations
        SCBA mounted in cab
            for quick donning
        CRT screens
            display incident information
        Global Positioning System
            allows vehicle tracking
            in rural areas allows relay of exact position to aircraft
        double battery set up
            for increased starting power
            heavy amperage draw of multiple warning lights

## Apparatus Motor

        diesel becoming very common
            long life and durability
            abundance of torque
            turbo charged and/or supercharged
            oversized alternator
            high idle switch
                keeps amperage up

### Modular Apparatus

    cab and chassis
        replaceable build up
            one cab and chassis for various specialized uses

### Pumper/Engine

    triple combination most common
        hose, water tank, and pump
    water tank
        vary in size
            commonly 200 to 1,500 gallons
        baffles to prevent weight shift caused by sloshing
        plastic gaining in popularity for tanks
            non-corroding

### Pumps

    main purpose is to lift water
    centrifugal pumps
        most commonly used as main fire pump
        *Transparency 17
            vaned wheels inside casing called impellers
            suction inlets introduce incoming water into eye
            impeller spins, forcing water towards outside of casing
            casing uses volute area to convert centrifugal energy to pressure
            water discharged at volute into plumbing
        advantages
            can spin without discharging water, allowing nozzles to be opened and closed

*if no water is discharged it can become quite hot and damage the pump or scald personnel!*

            takes advantage of pressure from inlet side
            can pump dirty water
            equipped with screens to limit size of debris entering the pump
    relief valve
        reroutes excess pressure back into suction side

*provided for firefighter safety, should always be set at safe operating pressure!*

        reduces pressure surges
    disadvantages
        can only act on the water that enters

cannot draw water from a static source

if it exceeds the intake pressure, it can cavitate and damage pump

### Positive Displacement Pumps

several forms
- *Transparency 18
- rotary gear, piston, diaphragm
- most common on fire apparatus is rotary gear

discharge volume equals intake volume

advantage
- self-priming, can pump air

used to "prime" centrifugal pumps
- commonly called "priming pumps"
    - evacuates air from pump casing, creating vacuum inside centrifugal pump
    - draws water from static source into centrifugal pump
    - allows pumpers to operate at draft
    - also can be used in some applications for high pressures

disadvantages
- if discharge is shut down, pressure builds up to point where something may explode
- cannot pump debris
- heavy and expensive in contrast to centrifugal pumps

### Aerial Ladder and Elevated Platform Apparatus

two basic configurations
- tractor/ trailer with tiller
- straight chassis

ladder types
- extendable with flies
    - with or without basket/elevated platform
- articulated boom
- 108 feet of ground ladders carried to qualify as ladder truck
- intercom system to provide communications between tip and base
- some have air system in basket for breathing air
- some plumbed with piping for elevated stream/water tower operations

quint
- ladder truck/pumper combination
- pump, water tank, ground ladders, hose bed, and aerial

squads
- designed for special purposes
    - lighting, air supply, emergency medical, hazardous materials, rescue, salvage, command post, tactical support, rehab, and so on

### Aircraft Rescue Fire Fighting Apparatus (ARFF)

    specially designed for aircraft fire fighting
- large water tanks
- foam and dry chemical systems carried (twinned system)
    - may be two nozzles hooked together
- often all-wheel drive for on and off runway use
- turret and ground sweep nozzles
    - remotely operated from cab
- pump and roll capability
- hand lines and ladders for access to fuselage

### Fire Tools and Appliances

- hose
    - used to get water from source of supply to seat of fire
    - construction types
        - hard rubber, cotton jacket, and synthetic
            - cotton jacket requires extensive maintenance compared to other types
            - must be washed, and dried inside and out
    - couplings
        - to connect hoses together
        - male and female ends
        - different materials used
            - brass or pyrolite
            - threaded or quick connect (Storz)
    - carried in hose bed on apparatus
    - various hose sizes from 3/4-inch to 6-inch and larger
    - attack lines
        - small enough to placed into operation by one person
        - large enough for sufficient water flow
            - usually 1½-inch or 1¾-inch with 1½-inch couplings
    - supply hose
        - can be laid as apparatus drives forward
        - 2½-inch or larger
        - 2½-inch can also be used as a large volume attack line
        - large diameter hose (LDH), 4-inch plus, requires less pressure to move large volumes of water
    - wildland
        - 1-inch and 1½-inch used
        - reduced weight for advancing hose lays in rough terrain
        - rolled or bundled (hose packs) for carry
    - hard rubber (red line/ booster line)
        - carried mounted on a reel

easy to use and reroll
low volume
suction hose
for drafting
wire wrapped to resist collapsing under vacuum

### Nozzles

numerous sizes and designs
combination nozzles can be adjusted for straight stream or fog patterns
adjustable flow rate
bail for turning on and off

### *designed to turn itself off if dropped!*

hand held operate up to around 250 gpm
large bore up to 2,000 gpm
combination have adjustable flow rate and stream pattern
straight tips vary in size
greater reach and penetration
monitors are removable for placement
for limited access or safety of personnel
can operate unmanned
foam inductor
uses venturi principle to introduce foam concentrate into hose stream

### *manufacturer's instructions must be followed or quality foam will not be produced!*

### Fittings

for connecting hose and appliances
allow versatility
reversing direction of hose lay (male to male, female to female)
double male and double female
changing hose size or thread in lay
reducer and or adapter
female size mentioned first ( i.e. 2½-inch by 1½-inch)
combine or divide hose lays
wye and siamese

### Ladders

### *must be designed for firefighting use!*

attic ladder
10 foot, folding
roof ladder
14 foot with hooks

extension
> various lengths

### Self-Contained Breathing Apparatus

allow entry into hazardous atmospheres
> low oxygen concentration
> heated gases
> smoke
> toxic gases

consist of a face mask, air bottle, backpack, and regulator

***regulator operates in positive pressure mode to prevent entry of hazardous atmosphere into mask!***

### Hand Tools

general and specialized tools carried
many adapted from other uses
> hose tools are called spanners
> hydraulic rescue tools
>> force open car doors, lift dash boards, cut off roofs and brake pedals
> air bags for heavy lifting
> axes, circular saws, chain saws, and sledge hammers for opening structure roofs
> pike poles and rubbish hooks for removing material
> power fan
>> allows positive pressure ventilation
>> option to firefighters on the roof over the fire

***lightweight construction is collapsing sooner, making roof ventilation more dangerous for firefighters!***

> salvage covers
>> used to cover room contents to prevent water damage

*Note:* Departments are carrying foam blocks or using canned goods to raise furniture above water on the floor caused by suppression operations

>> used to gather material when pulling ceilings
>> water chutes and sumps

fire extinguishers
> both dry chemical and water

medical aid gear
> resuscitator
>> used to administer oxygen and assist breathing
> medical aid kit
>> bandages, splints, blood pressure cuff, and other gear

electrical generator
> for powering lights and tools

### Wildland Fire Fighting Tools

specially designed for wildland fires

McLeod, Pulaski, axe, shovel, chain saw, fusee, flagging tape, canteen, portable pumps and tanks

*cutting tools are kept very sharp and great care must be exercised when using them!*

### Heavy Equipment

bulldozers used in wildland fire fighting
- create fire lines
- some equipped with tank and pump (Pumper Cat)

special foam units for oil fire fighting

### Personal Protective Equipment (PPE)

modern station uniform
- flame resistant fabric
- wear cotton underwear

*any material made from nylon or related synthetic fibers has the possibility of melting to your skin!*

additional layers help to protect from radiated heat

structure fire PPE
- designed as a system
  - all of it is to be worn together
- consists of helmet with ear protection, SCBA, coat, pants, boots, and gloves

### Personal Alarm/Personal Alert Safety System

carried when entering hazardous situations
- sound alarm if turned to on position or you lie still

### Proximity Suit

used in ARFF
- can be used to approach fire, not for walking through flames

### Wildland PPE

for wildland fire fighting
- consists of hard hat with ear and face protection, goggles, fire shirt, flame resistant pants, lace up boots, gloves, and fire shelter

*Anytime you are operating on a wildland incident you should have your fire shelter with you!*

### Emergency Medical PPE

for use at EMS incidents

consists of disposable long-sleeved shirt of moisture resistant material, latex or vinyl gloves, eye and nose/mouth protection

### Aircraft

fixed wing (airplanes)

lead planes for recon and guiding air tankers

air tankers for applying fire retardant

transport planes for smoke jumpers, supplies, and crew transport

rotary wing (helicopters)

for recon, water dropping, and crew and equipment transport

emergency medical transport

This list of items used by firefighters to perform their various jobs is not nearly a complete one. It only serves as an introduction to some of the tools and devices in use.

## Review Questions   (answers appear in bold italics)

1. List several advantages of a fire department having its headquarters separate from a fire station. ***This places the administrative staff away from the noise and activity of a working fire station. It also separates the staff from line personnel and avoids problems caused by conversations overheard accidentally or on purpose.***

2. List three structures that may be located at the training facility and their uses.
    1. ***Burn building: used for live fire training in a structure.***
    2. ***Drill tower: used for ladder operations, high-rise simulations, and high angle rescue.***
    3. ***Classrooms: used for indoor instruction.***
    4. ***Apparatus room: used for storage of materials and as a classroom when needed.***
    5. ***Drafting pit: used for apparatus and operator testing.***
    6. ***Various props: used for rescue, extrication, hazmat and other types of training.***

3. What is meant by an enhanced 911 system? ***The system shows the address from which the call is received. Some have the capability of showing which equipment is to be dispatched and cross streets to aid in location of the incident.***

4. Why is the diesel motor chosen for most pumper apparatus? ***Because of its long service life and high torque ratings. It is the motor of choice for heavy trucks.***

5. Explain the difference between centrifugal and positive displacement pumps. ***Centrifugal pumps can only act upon the water that enters them. They impart energy to the water through developing centrifugal force. They provide a certain amount of slippage that allows the flow to be started and stopped without adjusting pump speed. A positive displacement pump discharges a specified volume of water for each revolution or stroke. If the flow should stop, the pump discharge would need to be redirected or the pump turned off to prevent over pressure and rupture of the hose.***

6. Why is a centrifugal pumper equipped with a positive displacement pump in conjunction with the main pump? ***To evacuate the air and allow the pump to draft water from a static source.***

7. List three types of squad vehicles and their functions.
    1. ***Rescue: either an ambulance or vehicle equipped with rescue equipment.***

2. *Air unit: equipped with a compressor or cascade system for refilling SCBA bottles. Some departments forego the refill capability and transport full air bottles in this vehicle.*
3. *Rehab unit: used to carry drinks and provide shade or a heated/air conditioned environment to rehabilitate firefighters at the scene.*
4. *Hazardous materials: used to transport the specialized equipment used by hazardous materials teams.*

8. What is meant by a "twinned system" on aircraft rescue fire fighting (ARFF) apparatus? *This system is a combination of AFFF (aqueous film forming foam) or other foam and a dry chemical extinguishing agent.*

9. What is the difference between attack and supply hose lines? Give examples of each. *Attack hose lines usually range in size from 1 to 2½-inches in diameter and are advanced to attack the fire. Supply lines usually range in size from 2½-inch and larger in diameter and are used to supply water to locations remote from the water source.*

10. List the four main components of a self-contained breathing apparatus (SCBA). *Mask, backpack, air bottle, and regulator.*

11. What are the uses of a resuscitator? *To provide oxygen to patients, with some equipped with a suction device to clear foreign matter from the patient's airway.*

12. List three wildland fire fighting tools and their uses.
    1. *McLeod: scraping and raking.*
    2. *Pulaski: chopping and digging.*
    3. *Shovel: digging, scraping, throwing dirt, and limbing small trees.*
    4. *Combi: the same uses as a shovel.*
    5. *Chain saw: cutting brush and trees.*
    6. *Drip torch and fusee: lighting fires for backfiring or burning out.*

13. What are the components of structural personal protective equipment (PPE)? *Self-contained breathing apparatus, helmet, coat, pants, gloves, and boots.*

14. What are the components of wildland PPE? *Hard hat, goggles, fire shirt, pants, gloves, boots, and fire shelter.*

15. What are the components of EMS PPE? *Eye, mouth/nose protection, long-sleeved liquid-proof garment, and rubber (latex) gloves.*

16. What fits the description of fixed wing fire fighting aircraft? Give two examples of their uses. *Airplanes—they are used as air tankers, lead planes, and transports.*

17. What fits the description of rotary wing fire fighting aircraft? Give two examples of their uses. *Helicopters—they are used for water drops, and for crew and equipment transport.*

# Fire Department Administration

## Learning Objectives

*Upon completion of this chapter, you should be able to:*

- Describe the six principles of command.
- List and describe the six components of the management cycle.
- Identify the four methods of communication.
- Describe fire department chain of command.
- Fill out a typical fire department organizational chart.
- Identify different fire department types.
- Identify the different ranks and their general responsibilities.
- Explain the terms customer service, one department concept, team building, and incident effectiveness.

## INTRODUCTION

To effectively participate in the fire department, it is necessary to understand how it operates. In this lesson you will be introduced to the organizational structure of the fire department and how it is managed.

## LESSON PLAN

### Administration

    one of the most important jobs
    chief balances needs of community and department with resources available

### Principles of Command

    unity of command
        one boss concept
        violated at emergencies
            conflicting orders
    chain of command
        interlinked levels of authority and responsibility
        for new firefighter, chain in effect stops at company officer
            unless special assignment
            as rank increases chain extends upward
        formal path of communication
        allows command structure to maintain control
        prevents "going around" or "over head" of supervisor
        informal communication system
            grapevine or rumor mill
            information may be untrue
            based on speculation
            can hurt careers
            gossip
        chain of command should be strictly adhered to
            violated only in extreme circumstances or emergencies
        applies to unity of command
        who answers to whom
        organization chart does not specify all relationships
            even on chart does not mean even in fact
                rookie firefighter in contrast with veteran
        small department has simple organization chart
        large department has complicated chart
            numerous bureaus and divisions
            at top of chart is the chief
                ultimate responsibility for department
        transfer of command at incidents
            ranking officer has choice of taking over

- - - frees up company officers
- - - - advanced training in large incident management
- - on small incidents may just observe
- - as incident winds down command passed down
- - - company officer responsible for termination of incident
- - organizations divided into line and staff
- - - line: directly related to goals and objectives
- - - - preservation of life and property
- - - - - through prevention, suppression, and rescue
- - - staff: support line functions
- - clear division blurs in area of prevention
- - - considered staff
- - - do not perform suppression
- - - directly related to mission
- span of control
- - supervisor can only control so many subordinates effectively
- - - nature or complexity of work a factor
- - - - in fire fighting 3 to 7 considered optimum
- - - - situation is dynamic and dangerous
- - - - excessive division inhibits coordinated attack
- - commander needs to know what is happening
- - - does not need to know every detail
- division of labor
- - work divided to prevent duplication of effort
- - to apply most appropriate resources
- - based on area, skill, and complexity
- - seen in division of line and staff
- - - staff does not respond to emergencies
- - - increases productivity
- - - - less disruptions
- - division by area of responsibility
- - - prevention does plan check
- - - arson investigates suspicious fires
- - - allows specialization
- - fire stations located for effective coverage of risk
- - - concentrated in high-value districts
- - - less concentrated in rural areas
- - first-in company has local knowledge
- - assisting companies access knowledge of first-in officer
- - better than one central fire station
- - - shorter response time
- - - increased knowledge of own first-in area
- - specialization by function
- - - saves money and time

hazbmat team highly trained
    no need to train everyone in the department to same level
delegation of authority
    delegate to get work done
    frees time for planning and managing
    supervisor responsible for seeing that subordinates adequately trained to perform required jobs
    must give right and responsibility to make decisions (empower)
at emergency scene
    many jobs to be performed simultaneously
    delegate through assignments of functional responsibility
        suppression
        rescue
        treatment
        transport
commander cannot make every decision
    overload leads to inaction
authority can be delegated
responsibility cannot
    still responsible for actions of subordinates
    requires training of subordinates in proper methods
exception principle
    notify superior in situations of major importance
        unusual circumstances
        personnel matters
        major expense to department
        major incidents
    situation may require notification of superior
        clarification of policy
        another perspective

## The Management Cycle

*Transparency 19
organized thought process to achieve the desired goals of the organization
    goal: a general statement of a desired result
        broad in nature
        non-specific
        tend to be unmeasurable
    illustration: "Provide effective life safety"
        objectives
            define goals
            measurable
            attainable

- understandable
- clearly defined
- illustration: "Reduce fire deaths to one per ten thousand population"
- planning
  - the first step
    - determine objectives
    - decide on resources
    - requires determination of policies
      - defined: definite course or method of action
        - tend to be broad, encompassing variety of situations
        - should be written and clearly understood
    - procedures: specific way of doing a job
      - address situations covered by policies
- organizing
  - second step
  - incorporate resources in a structured relationship
    - organization table
    - chain of command
  - structure designed to address needs determined in planning
- staffing
  - third step
  - assignment of resources
    - to carry out plan
    - placed in pre-determined structure
  - debate over staffing level for fire companies
    - study conclusions
      - more personnel, more work in allotted time
      - more complex tasks require more personnel to complete
      - working above or below grade wears personnel out more rapidly
      - ambient temperature is a critical factor
      - quality of leadership makes a great difference
      - level of training, physical fitness, and competence is a major factor
    - must adequately staff both line and staff functions for efficient operation
- directing
  - fourth step
  - guiding and supervising subordinates toward objectives
  - accomplished through
    - rules
    - standard operating procedures
    - job descriptions
    - assigned duties
  - rules set limits for behavior
    - must be written

what is expected of employee
when discipline is required
job description
specifies duties and responsibilities
specifies minimum requirements
"and other duties as assigned"
to validate the testing process
testing matches level of job
controlling
fifth step
keeping department on track
to achieve goals and objectives
budget is a control tool
financial officer updates chief on state-of-budget
evaluating
determining if goals and objectives are being met
measures department against own goals and objectives
carried out internally and externally
internally
progress reports
*Transparency 20
station inspections
program records
fire losses
externally
feedback from public and elected officials
questionnaires

*Note:* Some departments have started sending questionnaires to citizens they have served on incidents to see how they performed in the eyes of the customer.

evaluation must use accepted standards
not all departments are equal
staffing
response times
employee evaluation
*Transparency 21
on-going process
evaluates strengths and weaknesses

## Fire Department Types

volunteer
first fire departments in U.S. were volunteer
90% still entirely or mostly volunteer

- protect 42% of the population
- participate in fund-raising to buy equipment
- training programs available from NFA
- no personnel costs leaves money for equipment
- *Transparency 22
- most common paid position is driver/operator
    - maintains equipment
    - brings it to fires
- pagers used to alert and dispatch volunteers

combination
- *Transparency 23
- part paid and part volunteer
- usually paid officers and driver/operators supplemented by volunteer firefighters
- saves on personnel costs when large staff rarely needed
    - allows for trained leadership
- common practice is to call volunteers reserves
    - reserves receive training and experience
    - jurisdiction receives trained force at little cost
- state- and federal-mandated training jeopardizing concept
    - untrained personnel can lead to legal action against department

public safety department
- *Transparency 24
- police and fire combined
- handle both types of calls
- cannot overly specialize due to demands of both professions

paid fire department
- prevalent in large cities
- *Transparency 25
- required level of service demands full-time professionals
- jurisdiction has more control over personnel
    - can hire and fire
- requires expert management
- budgets in the millions of dollars
- firefighting background may not determine whether a person is hired
- qualifications include
    - advanced college degrees
    - Executive Fire Officer
- requires ability to communicate
    - verbal and written skills
    - little direct contact with lower ranks
- popular management concepts
    - customer service

both internal and external
- one-department concept
    - standardization of equipment and procedures
- team building
    - working together within department guidelines
    - common goals
    - looking out for each other
    - accepting responsibility to help out others
- incident effectiveness
    - ability to act quickly and efficiently
    - based on drill, equipment maintenance, programs, and station maintenance
- rank structure
    - chief
        - overall leader
    - deputy/assistant chiefs
        - functional or geographic areas
        - directly assist chief in administration of department
    - battalion/district chiefs
        - several fire stations or major functions
    - company officers
        - one station
        - first level supervisor
        - supervises station personnel
    - driver/operator
        - operates equipment
    - firefighter
        - does whatever he is told to do
- industrial fire brigades
    - organized to protect specific plant or process
    - made up of plant personnel
        - training from employer, private, fire department, or combination
- contract fire protection
    - provide suppression, prevention, medical aid, and hazmat
    - by contract or subscription
    - if not a subscriber, pay suppression costs
- generate cost savings by
    - reduced personnel costs through paid reserves, lower salaries and benefits
    - utilize standby time for other activities
    - use of innovative strategies and technologies
    - public fire departments are starting to do the same thing as well

### Communications
 four basic methods
  face-to-face
   preferred method for transferring command at incidents
   provides most feedback
  radio
   quick when personnel are separated
   has limitations
   technology is improving
   cellular phones gaining popularity
   easy to listen in on with scanner
   be professional
   keep messages short and for essential business
  written
   used when not time critical
   used to maintain a hard copy
  electronic
   fax and modem
   provide hard copy when desired
   can link remote sites
   reduce volume of paperwork
   speed up delivery of information

## Review Questions (answers appear in bold italics)

1. Summarize the six principles of command.
   1. *Unity of command: everyone answers to only one person.*
   2. *Chain of command: all communication and decision making travels in an orderly fashion up and down an interlinked chain.*
   3. *Span of control: the organization is set up in a way that people do not supervise more personnel than they can manage effectively, usually 3 to 7.*
   4. *Division of labor: the jobs to be done are divided into manageable workloads and allow for specialization where necessary.*
   5. *Delegation of authority: when personnel are assigned a task they are empowered with the authority as well as the responsibility to see that it is completed.*
   6. *Exception principle: the supervisor is advised of situations or occurrences that are out of the ordinary.*

2. List the six components of the management cycle. *Planning, organizing, staffing, directing, controlling, and evaluating.*

3. Why is it necessary for the management cycle to be a continuing process? *New demands are constantly being placed on the organization. As these occur, the process must be applied to satisfy the new demands and evaluate progress.*

4. List the four forms of communication and their strengths and weaknesses.
   1. *Face-to-face: the best form of communication, as it provides for feedback from both parties.*
   2. *Radio/telephone: allows for rapid communication over distance, but provides limited feedback.*

3. *Written: messages can be sent with great clarity and provide a record, but the process is often too slow.*

4. *Electronic: has the advantages of the written message with the additional advantage of speed in transmission, but limited access to the necessary equipment is often a problem.*

5. Which form of communication is the most effective? Why? *Face-to-face, because you can observe the other person's face and they can see yours. Through body language, we can often discern whether the other person really understands what we are talking about. At incidents, you can point to a location or object and the other person can see it.*

6. In the chain of command, who does the firefighter answer to? *The company officer.*

7. Draw and fill in the blanks on a simple organization chart with at least five levels of rank represented. *This chart should begin with the chief at the top, and by using various titles proceed down through company officer to firefighter.*

8. What are the general responsibilities of the firefighter? *To do as ordered in a safe manner and report back when the task is completed.*

9. What are the general administrative responsibilities of the company officer? *The company officer's responsibilities require that the company be trained, equipped, and prepared to perform the functions required of it either on emergencies or routine assignments. There are numerous tasks that can be listed as part of the answer to this question.*

10. Who is the first level supervisor in the fire department? *The company officer.*

11. If you had a question about policy or procedure, who should you ask? *The first thing to do, if there is time, is to look in the department's manual of operations and then discuss the findings with the company officer. If time is not available, ask the company officer.*

12. What is the difference between line and staff functions? *Line functions are the personnel assigned to the fire fighting equipment. Staff functions provide support in the form of training, prevention, dispatch, and supply, as well as administrative functions.*

13. What is meant by the term "incident effectiveness"? *The ability of the department's personnel to perform their functions safely and effectively at incidents. This requires physical fitness, training, and equipment.*

14. Why is customer service such an important concept in the modern fire service? *The public expects service from the people who work for them. To serve the public as well as the customers inside the department, the personnel must be capable of performing their jobs efficiently.*

15. When issued conflicting orders by supervisors, what should you do? *Advise the supervisor giving you the conflicting order that you are already assigned and ask him to advise the other supervisor.*

16. Give an example of division of labor in the structure of the fire department.
*The answer to this question can be any of the specialty positions of line and staff.*

# Support Functions

## *Learning Objectives*

*Upon completion of this chapter, you should be able to:*

- Identify the support functions required by the fire department.
- Describe the duties and responsibilities of the support functions.
- Explain the need for the support functions.
- Explain the difference between a managerial support function and a technical support function.

## INTRODUCTION

An organization as complex as the modern fire department requires a long list of support personnel to perform its mission. By understanding these positions and their responsibilities, we can better perform within the framework of the department. It is always helpful to know where to go when you need help.

## LESSON PLAN

There are numerous positions in the fire department that perform support activities for the personnel in the fire stations.

### Dispatch

    receive and transmit requests for service
    may be firefighters or civilians
    firefighters familiar with needs of operations personnel
        can anticipate needs
        tend to be action oriented
        no outlet for adrenaline
        feeling of wanting to help
        many do not want to spend time in this position
    expanded dispatch
    used on large incidents
    takes load off regular dispatchers
    location remote from dispatch center
        Example: South Ops
            tracks and assigns resources on a regional basis
        NIFC on a national basis
            tracks daily availability
            helicopters, air tankers, crews, etc.
            closest available resource assigned
            regardless of jurisdiction
    transmission of alarms
        several methods
    911 and alarm pull box
        pull box still used in some areas
            does not identifyy type of problem
            dispatch sends "full box"
                false alarms
                vandalism and theft
            easy to understand and use
    proliferation of telephones
        911 system
        two-way communication
        determine appropriate response

                medical aid, fire, etc.
                    medical instructions from dispatcher
                        keeps dispatchers on phone
            some do not understand system
            language barrier
        lookouts
            sat in tower all day
            could not touch anything during lightning storms
            air pollution prevents clear view

### Graphic Arts/Maps

    keeps maps current
    GIS system
    creates overlays for maps
        occupancy and occupant information
        location information
        latitude and longitude for aircraft response
    GPS for mapping fires
    determines jurisdiction and property owners

### Hazardous Materials Control Unit

    formed due to state "right to know" laws
    regulation and control of hazardous materials
        storage and use
            business plans and inventory
                name and amount of materials on site
            information for emergency responders
            technical expertise
            inspection of storage and handling

### Arson Unit

    heinous crime
    no way to tell who will be injured or killed
    reasons for arson
        spite/revenge
        profit
        sexual gratification
        cover up other crimes
    arson discovered but hard to prove responsible party
    tends to destroy evidence
    arson bureau
        determines cause of suspicious fires

    not needed at every fire scene
    called when crime is suspected
    serious injury or death
    internal affairs bureau
      investigates complaints against employees
  arson investigators
    specially trained in gathering and preserving evidence
    interviewing suspects and witnesses
    fire fighting background an asset
      asked opinion in court
    need to know
      chemistry and physics of fire
    must be systematic and careful in their approach to scene
    good presentation skills for testifying and report writing
  cost recovery
    through billing and civil suits
    to recover expenses after incidents
      criminal or negligence cause
  extraordinary hazard ordinance
    recover costs incurred when extra hazard causes incident

## Personnel

  department personnel clerk
    matters regarding employees
      payroll
      insurance
      hiring and firing paperwork
      generates and retains evaluations
      assists with preparation of budget
  jurisdiction personnel office (Human Resources)
    entrance examinations
    promotional testing
    affirmative action and equal employment opportunity matters
    assures federal and state guidelines are met
    maintains currency of job descriptions
      may justify reclassification or pay increase
      change in hiring requirements

## Information Systems

  departments have almost too much information
  needs to be collected, organized, and retrieved
    incident information

apparatus information
personnel information
budget information, etc.
systems analysts
repair and service computer equipment
software and hardware management

### Business Manager

chief financial officer
budget
preparation and monitoring
accounting
income and expenditures
compliance with laws
fire business management
system set up before incident
contracts and resource information
requires inspection of equipment prior to use
accounts for supplies delivered to incident
personnel and equipment time for payment
technical support
legal services
contracts
mutual and automatic aid agreements
liability and workmen's compensation issues
crime lab services
through police department
National Weather Service
weather specialist for incident
spot weather forecast
hazmat incidents
health department personnel
data bases and books
special hazards
subways
refineries
chemical warehouses, etc.
EMS
training

### Warehouse/Central Stores

storage of often required items

receive supply requisitions from stations

fill orders

repair items

hose

SCBA

turnouts, etc.

### Repair Garage

services fire equipment

mechanics specially trained in fire apparatus

### Radio Shop

repair of radio equipment

HT's, base stations, vehicle mounted

### Adjutant/Aide

chauffeurs chief

assists with paperwork

assists on assignments

opportunity to learn about department on higher level

## *Review Questions* (answers appear in bold italics)

1. List three reasons why a fire department requires a dispatch center.
   1. ***To receive calls for service.***
   2. ***To dispatch calls.***
   3. ***To track resources and their availability.***
2. What is meant by the term *expanded dispatch*? ***Expanded dispatch is a system wherein another location is used to handle radio traffic and request resources for a particular incident or set of incidents.***
3. Why is there a need for expanded dispatch on major incidents? ***Expanded dispatch is set up when incidents reach so great a magnitude that the normal dispatching services are no longer capable of handling them along with their regular traffic.***
4. How are alarms transmitted to the dispatch center? ***There are several methods: alarm company tie-ins, phone calls, radio calls, and alarm boxes. In some cases, the dispatch center may receive requests for service from walk-in traffic (still alarms).***
5. What are the advantages of the GIS system? ***The GIS system allows for a system of map overlays to be used to assign names and coordinates to locations in the system.***
6. What valuable information can the hazardous materials control unit provide to operations personnel at the scene of an emergency involving hazardous materials? ***In a jurisdiction that maintains business plan files, the hazardous materials control unit has inventories of the materials on-site at different locations.***
7. Other than investigating fire cause, what are the functions of the arson unit? ***The arson unit personnel are often involved in Internal Affairs investigations.***
8. Why is it necessary to assign a personnel clerk to the fire department? ***The fire department in a large jurisdiction can have a large number of employees. The assigning of a personnel clerk to the department allows for a person who specializes in fire***

department issues and needs to take care of personnel matters.

9. As fire departments become more computer dependent, how will the role of information systems expand? *As the computers spread to the outlying stations, the wide area network and its hardware must be maintained and managed. New software must be purchased and maintained to increase efficiency and ease of use. Personnel must be trained in the use of computers for them to be fully utilized.*

10. What functions does the business manager perform when dealing with the department's budget? *The business manager ensures that all laws that apply are followed. He also maintains records on money as it comes in and out of the department. Bills have to be paid and the current state of the budget must be available because it directly affects decision making. The business manager also assists in statistical analysis for planning purposes.*

11. Why is a business management (finance) section set up on major incidents? *Large incidents often cross jurisdictional boundaries and each department's share of the cost must be determined. When equipment and personnel are contracted to the incident, financial records for personnel and equipment time must be maintained so the proper amounts can be paid. Any claims for damage to property or equipment must be processed and paid.*

12. Why is a weather person important during major incidents? *Weather is one of the most changeable factors during any incident. Numerous firefighters have lost their lives on wildland incidents because the weather changed and caused the fire to intensify or change direction. On hazardous materials incidents, weather can be a major factor in how vapor clouds will act and become a factor in determining if, when, and where evacuations are required.*

13. What are the advantages of a fire department having its own repair garage? *Personnel assigned to the department's garage can become specialists in the maintenance and repair of fire equipment. The department also has more input into which pieces of apparatus receive priority for return to service.*

14. If you were offered the position of chief's aide, what do you see as your responsibilities? *Your position would require you to drive the chief on his rounds and to assist with resource tracking on incidents. You could also be used as a runner to deliver messages.*

15. Your fire engine has a noise in the transmission and the radio is out of service. Who do you contact for service? *You would seek service at the department garage. In departments with a radio shop they would repair the radio.*

16. Can the fire department exist without the support services? *Justify your answer. No, the department requires the assistance of all of the support services to keep the apparatus in service, to dispatch calls, to gather and manage information, and to pay the bills.*

# Training

## Learning Objectives

*Upon completion of this chapter, you should be able to:*

- Identify the personnel and positions that make up a training bureau.
- Describe the need for training in the fire service.
- Explain the difference between technical and manipulative training.
- Describe how adequate level of training is determined.
- Describe how performance standards are determined.
- Explain how skills are developed.
- Explain the importance of skills maintenance.
- Explain how training level applies to incident effectiveness.
- List areas in which firefighters require training.

## INTRODUCTION

There are two main things required for firefighters to do their jobs. These are a willingness to participate and the training to perform safely, effectively, and efficiently. All of the most expensive fire equipment in the world is of no use if the people who are supposed to use it do not know how. Training can also provide us with the tools we need to make informed decisions when there are emergencies, as well as at other times. Many of these decisions will be critical to your survival and the survival of those around you.

### Lesson Plan

*many fatalities, in no matter what type of firefighting, have common denominators that have been recognized in previous fatality situations!*

    after firefighter fatalities occur, investigations are made
        conclusions are drawn
        recommendations are made
        reports are published
        made available on a wide scale

### Training Bureau

    teaches you to do your job safely and efficiently
    you are ultimately responsible for your own safety and education
    staff function
        40-hour, 5-day week
        may wear different uniform
        regarded as separate from operations
    operations function
        battalion training officer
        field personnel
        have aptitude and interest in training others
        develop programs to meet departmental needs
        vary with location
    personnel of training bureau
        able to plan
        plans must be flexible
            training interrupted by emergencies
        coordination of variety of resources
        research and develop information
        well developed communications skills
            verbal and written
        create visual aids
            video, slides, drawings, handouts
        present concepts and ideas
        must know material in depth

- to answer questions
- positive attitude towards the importance of training
- relationship to the organization
- wide background of experience
- may be specialty position
  - not awarded on seniority alone
- ability to change department
  - write policies
- work seen by all
  - have to be able to take criticism
- training chief
  - usually chief officer
    - lends authority to the position
    - one person in charge of training
      - consistency in output
    - presents a comprehensive program
    - contact point for upper ranks
    - prioritizes demands
    - if tied to college, acts as liaison officer
- company officers
  - act as training officers
    - present material
    - assist instructors
  - plan and coordinate programs
    - conduct research
    - make recommendations
  - evaluate others
    - needs assessment
  - may aid in promotion
  - greatly depend on cooperation of operations personnel
    - to deliver equipment
    - to assist when necessary
- instructors
  - may be from inside or outside department
  - specialists in presentation
  - made, not born
  - NFA instructors
  - teach all over the U.S.
  - considered experts
- A/V technician
  - plans and produces training visual aids
    - slide programs
    - videos

- can be economical in the long run
- keeps companies in their areas
- cuts down on mileage and other costs
- speeds up dissemination of information

interagency training
- those who operate together need to train together
    - other fire departments
    - industry
    - ambulance company
- learn each others' operations
- address problems before emergency incident
    - communications
    - equipment

training facilities
- need not be fancy
- can take place anywhere
- classrooms are a necessity
    - may be borrowed from a school or other facility
- drill tower
    - high-rise training
    - rappelling
    - high angle rescue
- might use parking structure
- props are expensive
    - utilize buildings under construction
    - facilities in the area
- sites require preparation
    - look for safety hazards
    - prepare to control fire if lit
- training during preplans
    - lay hose
    - predetermine ladder points
    - look for hazards
    - discuss possible operations and problems
- smoke generators
    - safe, clean smoke
    - can use to set up drills in ordinary buildings
- ARFF training
    - requires special knowledge
    - airport operations
    - communications with control tower
    - special equipment

### Purpose and Importance of Training

    purpose of instruction is to change behavior
        learn how to operate equipment
            wide variety used in fire fighting
        perform safely
        develop safety attitude

***there are many ways to perform most operations, but only a very few safe ways!***

    learn before required to perform
some jobs trained until they become habit
    donning SCBA
    size up
    development of thought processes so steps are not forgotten under pressure

### Technical Training

    training in facts and ideas
    not actions (manipulative)
    technical subjects
        chemistry
        hydraulic calculation
        building construction
        size up, etc.

***you must learn the design limitations of your equipment!***

    price for failure is too great
    exceeding design limitations can lead to injury or death

### Manipulative Training

    actual use of tools
    "perfect practice leads to perfect performance"
    under stress you will perform as you have practiced
    evolutions defined: performing drills with tools and equipment
    starts out very simple, one tool at a time
    as proficiency increases drills become more complicated
        proficient enough to work in the dark
        drills performed under realistic conditions
    stress drills used to test performance under heavy stress
        decision making
        physical fitness
    realistic within safety guidelines
    should only be used when personnel are proficient under normal conditions

### Determining Adequate Levels of Training

several criteria
- is the job being done safely?
- can the job be performed under realistic conditions?
- complexity of the job?
- how often is the job performed?
- how important is the job?
    - price for failure
        - injury or death
- is it mandated by law?

skills maintenance
- after mastering basics move on to more difficult
    - must maintain even low level skills

skills assessment
- testing
    - technical and manipulative
        - safety
        - all steps completed
        - efficiency
        - smoothness of operation
        - time to complete

### Standard Operating Procedures

standardize operations
- for unity and coordination

apply to areas of
- command
- communications
- safety
- tactical priorities
- utilization of companies

speeds up operations

cuts down on radio traffic

### Training Records

*Transparency 26

maintained on regular basis

list of training completed
- for legal reasons
- for planning

can be reviewed to assess deficiencies

for claiming reimbursement under apprenticeship programs

### Relationship of Training to Incident Effectiveness

- overall purpose of training is to achieve incident effectiveness
- operations performed with safety and efficiency
  - reduced staffing on equipment
- required training
  - public expectation to be able to handle anything
  - Federal OSHA 29 CFR
  - First Responder Operational level
    - personnel required to take action at a hazmat incident
      - identification of risks due to hazardous materials
      - potential outcomes of release
      - ability to recognize hazmat exits at emergency
      - ability to ID hazardous materials
      - ability to use DOT ERG
      - ability to realize need and call for additional resources
      - requires annual refresher training
  - Federal Aviation Administration Part 139
  - airport crash fire rescue training
    - airport and aircraft familiarization
    - personnel safety
    - emergency communications
    - equipment use
    - application of extinguishing agents
    - aircraft evacuation assistance
    - firefighting operations
    - adapting and using structure fire apparatus for aircraft fire and rescue
    - familiarization with duties under airport emergency plan
    - at least one live fire drill every 12 months
    - basic emergency medical care
  - EMT
    - requirements set by local or state level
  - firefighting skills maintenance
    - requirement by department
    - rule of thumb: 2 hours per person per day
    - may include preparation to assume position above you

## Review Questions (answers appear in bold italics)

1. Why is the training officer position given to a high ranking officer? ***The person in this position has responsibility for preparing and presenting training to all of the personnel in the department. With this responsibility must come the commensurate authority to carry out the tasks. If the training officer is a firefighter and the company officers choose to ignore him, the training will not be accomplished.***

2. Is the training bureau more of an operations or staff function? ***The training staff often works a staff schedule, but their purpose is to increase the efficiency and effectiveness of the operations personnel.***

3. Why do the members of the training bureau have such an effect on the direction of the fire department as a whole? ***In many cases they do the research and write the policies and procedures used by the department as a whole.***

4. Which level of government sets the requirements for first responder operations level training? ***The federal government.***

5. Which level of government sets the requirements for emergency medical service training? ***In most cases it is the state with local administration.***

6. Why is time alone not a good indicator of the performance level of an engine company? ***Because safety and efficiency are two important factors in any operation.***

7. Why is it important for firefighters to train with the local ambulance company? ***In most cases the firefighters will be working closely with the local ambulance service provider*** *at incidents.*

8. When there are multiple jurisdictions in the area, what are the benefits of joint training? ***Sooner or later an incident is going to require the joint efforts of the jurisdictions. That is not the time to find out you are not able to communicate or coordinate efforts.***

9. What are the two basic types of training a firefighter receives? ***Technical (book) and manipulative (motor skills).***

10. Which type is based on the operation of various tools? ***Manipulative, with some technical.***

11. Which type of training is learning firefighting-related chemistry? ***Technical.***

12. What is the importance of maintaining training records? ***Training records can point out deficiencies in areas where not enough training has occurred. They may also be subpoenaed into court when a lawsuit is filed.***

13. List two factors in determining an adequate level of training.
    1. ***Safety***
    2. ***Complexity of the job***
    3. ***Frequency with which the job is performed***

14. What is the one most important factor in any training? ***SAFETY FIRST, ALWAYS!***

15. Why are stress drills performed? ***To require personnel to perform under realistic conditions and to condition them to perform as required when things go wrong.***

16. What is the importance of standard operating procedures? ***They prescribe performance so that various companies can coordinate efforts at incidents. They also assist personnel in working with unfamiliar crews when reassigned or on overtime.***

# Fire Prevention

## *Learning Objectives*

*Upon completion of this chapter, you should be able to:*

- Describe the importance of fire prevention.
- Describe the activities performed by a fire prevention bureau.
- List methods of public education as it relates to fire prevention.
- Explain how the authority to enforce fire prevention regulations is derived.
- Describe a typical fire prevention bureau organization.
- Describe the importance of fire information reporting.
- List the uses of fire-related statistics.

## INTRODUCTION

One of the most important jobs the fire department performs is fire prevention. By stopping fires from starting, we are ahead of the game. The United States has one of the highest fire death rates per capita in the world. To reduce deaths, we must reduce the number of hostile fires that start. If no incidents occur, there will not be any victims or property loss. Fire departments are often judged by their fire loss experience.

## LESSON PLAN

### Fire Prevention Bureau

- prevention personnel often inspect the technical or high risk occupancies
- requires special training in fire prevention
- many are specialists due to knowledge required
- staff function
    - fire inspector
    - specially trained in
        - science of fire
        - fire prevention inspection
        - enforcement
    - maintains written records and reports
    - knowledge of
        - fire chemistry
        - building construction
        - electricity
        - safety practices
        - codes and ordinances
        - hazard recognition
    - not only identifies hazards and risks but works with property owner on correction
        - through education
        - suggestions on correction method
        - compliance through legal action
    - education and public relations very important
    - may be only contact owner has with fire department
- operations function
    - company level fire prevention work
        - inspection and education
        - need to know codes
        - must not overlook hazards
- personnel
    - in large department, supervisor may be called Fire Marshal
    - all must understand importance of prevention in achieving the mission
    - staff often mix of uniformed and civilian personnel
    - selected based on aptitude, attitude, and experience

Fire Prevention Chief
    in large department
        Chief rank due to complexity of operation
    small department
        company officer
    some departments
        civilian position
    recommendations to the Fire Chief
    politics involved in position
        some requirements are quite costly to business
    need to be well-spoken to sell program
inspectors
    variety of ranks
        company officers, firefighters, civilians
    technical specialists
        usually civilians
        hired for particular expertise

## Professional Standards

NFPA 1301, *Standard for Professional Qualifications for Fire Inspector*
training available on national, state, and local level from
    NFA
    colleges
    State Fire Marshal programs
    accompanying inspectors on inspections
subjects
    codes and ordinances
    plans review
    inspection principles
    public education
    instructional technique
prevention officer organizations
    contemporary issues
    networking
magazines and other publications
    issues and experiences

## Purpose of Fire Prevention Activities

prevent loss of life and property due to fire
    to prevent hostile fires from starting
    to provide for life safety
    to prevent the spread of fire from one area to another

### Fire Prevention Activities

four areas

the three E's: engineering, education and enforcement, and arson investigation

*Note:* Some personnel think the three E's are: extinguishers, exits and extension cords

activities include
1. design of fire safe assemblies and systems
2. review of plans prior to buildings being built or remodeled
3. inspection of fire safety equipment and devices once installed
4. inspection to ensure that the devices are kept in working order
5. enforcement of codes and ordinances related to fire prevention
6. public education in the methods and benefits of fire prevention and fire safety
7. education of the legislative body about the need for the enactment of fire safety-related legislation
8. investigation to determine fire cause and prosecution of arson when applicable

fire prevention takes place everywhere

overall goal is to keep people and property fire safe

most effective programs gain voluntary compliance
through education

### Fire Prevention Terms

inspection is the act of making a systematic and through examination of a premise or process to ensure compliance with fire codes and ordinances

hazards are anything that can cause harm to people or property

risks are the activities undertaken in relation to the hazard

occupancy is the use or intended use of a building, floor, or other part of a building
divided into types by use
A assembly
I institutional
E educational
S storage, etc.

### Methods of Fire Prevention

start before building is built
during design process
assemblies
devices
to protect special hazards
protection features
exits

zoning
location in relation to property lines
water supply for fire fighting (fire flow requirement)

***firefighters should be especially aware of lightweight construction because it fails early in fires!***

## Hazard Evaluation And Control

to identify possible accidents and estimate their frequency and consequences
- initiating event
- response of operators and equipment dictate subsequent events

evaluation and control through
- adherence to good practice
    - observing rules and regulations
    - use of accepted standards (NFPA, etc.)
    - following accepted procedures and practices
- deviation identified through use of
    - checklists and safety reviews

predictive hazard evaluation for analysis of
- processes, procedures, systems, and operations

where adherence to good practice may not be adequate

first step is to identify hazards

evaluated in terms of the risk it presents to
- people and property

evaluate events that could be associated with the hazards

various evaluation methods used (beyond scope of this text to present in depth)

## Public Education

to deliver the message to the most people

we cannot be everywhere at once

people must ensure their own safety

methods

school programs, station tours, community groups, PSA's, advertising

may get funding from private industry to purchase materials

involve the media
- get them interested in fire prevention
- through media day or other activities
- give them information they can use, especially on slow news days

## Organization

State Fire Marshal
- research, investigation, training, and prevention divisions

support local and act in their absence
local
- department prevention bureau
- more technical inspections and activities

company level
- general business inspections can serve several purposes

***by having firsthand knowledge of a building and its contents, fire fighting is safer and more efficient!***

- perform preplanning
  - locate utility shutoff
- perform public relations work
- learn where highest value is in occupancy
- hazard reduction/weed abatement
- educational opportunity for public
- identify extra hazards in first-in district
  - accumulations of tires and other items

## Fire Prevention Inspection

preliminary work
- review past inspections for previous problems
- identify special hazards and what is done to address them
  - systems and devices

have proper tools
- *Transparency 27
- uniform, hard hat, flashlight, tape measure, clipboard, forms, code book, etc.

codes allow entry when suspected hazards exist
- may need inspection or administrative warrant to gain entry
- more restrictive in some ways than a search warrant
- business license may specify entry is allowed

in most situations entry is welcomed
- announce your purpose, be polite
- may have to come at more convenient time for occupant
- have someone accompany you if possible

approach inspection in a consistent manner
- outside in, top to bottom, start of process to finish
- be thorough, look in closets and attics

prioritize hazards and violations
- life safety first (blocked exits)

***a blocked or locked exit should be corrected in your presence!***

- minor fire safety violation may be given a reasonable abatement period (out of date extinguisher)

record all violations
leave copy with manager or owner
thank them for their time and cooperation
in case of long list or serious violations
- develop plan of correction
- what must be done by when
- make reinspection at end of allotted time

if no compliance
- write citation

disagreement between inspector (code) and owner
- may go to board of appeals
- alternate materials and methods

### Determination of Fire Cause

chief charged with responsibility to determine cause on all fires
- delegates to subordinates

important for prevention reasons
- may identify where to target efforts in education, enforcement, and engineering

pay attention as you fight the fire

structure fire
- color of smoke and flames, position of doors, locks, difficulty of extinguishment

wildland fire
- area of origin and people leaving scene
- when fire is out, do not disturb area of origin if possible
- total overhaul may have to be delayed
- do not touch any probable evidence
- no statements to the press or anybody else who is not from the fire department

reconstruction is next step
- approach made with no preconceived ideas as to cause
- starts at area of least damage and moves toward area of origin
- witnesses interviewed, including firefighters

facts are determined
- fuel source, heat source, and act or omission that brought them together

conclusion is drawn

types of investigations
- basic, fires where cause is obvious
- technical, more in-depth
  - may require evaluation of evidence by crime lab
- incendiary, believed that a crime has been committed
  - usually conducted by arson unit
  - arson unit also called in cases of fire-related death or serious injury

to prevent missing criminal cause

may also be called if company officer is unable to determine cause

### Fire Information Reporting

*Transparency 28

fire reports are generated for several reasons

budget justification

trend analysis

needs assessment

identification of faulty equipment

justify code changes

identify need for enhanced safety devices

identify need for legislation

## Review Questions  (answers appear in bold italics)

1. List the four areas of fire prevention and give an example of each.
   1. *Education: school programs, service club visits, etc.*
   2. *Engineering: plan checks, system design and testing, etc.*
   3. *Enforcement: permits, inspection, weed abatement, etc.*
   4. *Investigation: cause determination, statistical analysis, etc.*

2. What is meant by a fire assembly? *A fire assembly is a device or combination of devices designed to suppress, retard the spread of, or prevent a fire from occurring.*

3. Give an example of a fire preventive device. *These are too numerous to list. They include fusible links, extinguishing systems, heat rise limiting switches, detectors, and so on.*

4. List two methods of presenting fire prevention education to the public.
   1. *School programs*
   2. *Visits to civic groups*
   3. *Passing out literature*
   4. *Public service announcements in the media*
   5. *Billboards*

5. You have been given the responsibility of preparing a fire prevention presentation for a third grade class. What are some of the things you can do to get your point across? *Pass out coloring books, speak to the class on their level, encourage them to visit the fire station, speak with the teacher, and have the students perform some activities before and after your visit to reinforce the lessons presented.*

6. List two advantages to a fire company performing fire prevention inspections in their first-in district.
   1. *Area familiarization*
   2. *Public education*
   3. *Public relations*

7. Why does the fire prevention bureau inspect the more complicated occupancies? *Because the level of knowledge needed to understand complicated fire systems is often beyond the training of the fire company.*

8. Explain the difference between a hazard and a risk. *A hazard is something that exists; a risk is an activity performed in relation to the hazard.*

9. What is the reason for weed abatement in an area with a dry climate? *Fires can easily spread from dry ground fuels to structures.*

10. What hazard do wood shake roofs pose from a fire standpoint? *They not only burn very well, but spread burning brands in their convection column.*

11. Illustrate the chain of authority that allows you to require compliance with the fire

code. ***This answer is going to depend on the jurisdiction involved. In most cases the code is adopted by ordinance and the Fire Chief is given the authority to enforce it The authority is then passed down through the organization to the members of the department.***

12. Which fire code is adopted in your area? ***Depends on the authority having jurisdiction and which code they have selected.***

13. Illustrate the organization of a standard fire prevention bureau. ***Fire Marshal, Assistant Fire Marshals, and Inspectors, along with plans check specialists and water systems specialists is one example.***

14. Where may a firefighter seek fire prevention training? ***From the National Fire Academy, state and local level training associations and several colleges across the country.***

15. When faced with an owner who fails to comply with the required corrections, what actions should you take? ***First, educate the owner as to the requirements and their necessity. Determine a plan of correction with the owner. If this does not work, a citation may have to be issued.***

16. List three reasons why fire statistics are collected.
    1. ***Budget justification***
    2. ***Trend analysis***
    3. ***Identification of need for new codes***
    4. ***Identification of faulty equipment***
    5. ***Needs assessment***
    6. ***Aid in determining strategies to overcome problems***

# Chapter 11

# Codes and Ordinances

## Learning Objectives

*Upon completion of this chapter, you should be able to:*

- Explain the relationship between federal, state, and local regulations.
- Explain who is responsible for enforcing codes and ordinances at the different levels.
- Explain why codes and ordinances are created.
- Describe how codes and ordinances are adopted.
- Describe how codes and ordinances are affected by court decision.
- Explain the relationship of codes and standards.
- Give the definition of legal terms as they apply to codes and ordinances.

## INTRODUCTION

Codes and ordinances fall under the definition of laws. Laws are written and adopted at all three levels of government: federal, state, and local. What the fire department and its members can and cannot do are governed by these codes and ordinances.

## LESSON PLAN

### Definition of Laws

- U.S. Constitution
    - supreme law of the land
    - other laws cannot conflict with it
    - statutory laws are passed by the Congress and state legislatures
        - called statutes
    - includes local ordinances
    - federal statutes organized into Code of Federal Regulations (CFR)
    - states do much the same thing (codes)
    - not to be confused with model codes (fire and building)
        - given force of law when adopted by local ordinance
    - constitutionality determined by judicial system
        - decisions referred to as precedents
    - when a law is not specific, current law and previous court decisions interpret intent
    - law makers cannot keep up with every new situation
    - need to know what the law is to know limits of authority and what is required
    - challenges may be decided in court or by board of appeals or chief

### Lawsuits

- lawsuits becoming very common
    - more attorneys than firefighters
    - attorneys and plaintiffs may receive money from outcome
- to avoid lawsuits, do your job correctly every time
    - document actions
- you are sued because of a tort
    - tort is a wrongful act
    - can result from
        - nonfeasance: failure to act
        - misfeasance: acting incorrectly
        - malfeasance: wrongdoing or misconduct
- to avoid torts, act within policies and guidelines that are:
    - agency specific and regularly reviewed for validity
    - basic responsibility is to perform correctly and to the best of your ability every time

## Personnel Complaints

prescribed procedure for persons wishing to file a complaint to follow
1. speak to chief or supervisor
2. officer explains options
3. form is filled out and forwarded to designated officer
4. investigation conducted, determination made
    was complaint warranted
    disciplinary action to be taken, if any
5. person complaining notified as to results of investigation
    warranted or not
    not advised as to disciplinary action taken

## The Court System

concept of jurisdiction
    limit of territory within which authority may be exercised
        territory may be functional or physical
        fire prevention versus speeding
        county versus city
    court of original jurisdiction is where case is heard first
    appellate jurisdiction reviews lower court determination
    highest court is the U.S. Supreme Court
        constitutionality of laws
    federal district courts hear violations of federal law
        illegal campfire in Forest Service jurisdiction
    state supreme court hears appeals from district courts of appeal
    state law violations heard in district or superior courts
        arson, fireworks violations, etc.
    municipal or county courts
        misdemeanors; illegal trash burning, parking in fire lane, etc.
    must know jurisdictional boundaries
        physical and functional
        cannot legally act outside of boundaries
    may be able to handle through referral to department with jurisdiction
        accumulation of trash referred to health department
        open, vacant building referred to building department

## Fire Prevention

legal responsibility and authority to enforce fire-related codes and ordinances
    authority to inspect quoted under *See vs. City of Seattle*
    U.S. Supreme Court set forth guidelines

1. inspector must be adequately identified
2. must state reason for inspection
3. must request permission to inspect
4. invite person to accompany you
5. carry and follow written inspection procedure (inspection form)
6. request inspection or administrative warrant if entry denied
7. may issue stop order for extremely hazardous condition
8. develop reliable record keeping system
9. work within guidelines
10. must have right to inspect, may be through licensing
11. inspectors must be trained

most local codes are national codes adopted by ordinance
    in part or in whole
    amended as necessary

state fire marshal may enforce in:
    state buildings
    areas with no organized fire prevention
    often delegated to local authority

### Model Fire Codes

BOCA: *Basic Fire Prevention Code*
ICBO: *Uniform Fire Code*
SBCCI: *Standard Fire Prevention Code*
NFPA: NFPA 1 *Fire Prevention Code*

*Note:* At this time (July 1996) three of the model code bodies (ICBO, SBCCI and BOCA) have formed the International Code Commission to formulate an international fire code that will standardize the model fire codes they publish. With the purpose of having the code serve as the model adopted throughout the U.S., the NFPA is promoting NFPA 1 as the code that should be adopted.

model codes are desirable in that they are:
    nationally recognized
    based on research and development on a large scale
    designed to not conflict with building codes

codes and standards
    codes are created to be adopted as law
        by ordinance
    standards are designed to serve as recommendations
        NFPA 1901 *Pumper Fire Apparatus*
    standards are sometimes adopted by memorandum of understanding
        NFPA 1500 *Fire Department Occupational Safety and Health Program*

### Occupancy Classification

when building is built, occupancy must be identified
    determines whether certain parts of fire code apply

determines fire and life safety features to be included
occupancy classifications (partial list)
- A assembly
- B business
- E educational
- I institutional
- R residential

many classifications have subcategories to address specific problems

after construction, if occupancy changes requirements may be added
- sprinkler system retrofit for higher hazard occupancy

building code requirements based on occupancy
- construction components, fire resistiveness
- area and height
- set back from property line
- fire protection systems installed

## Construction Types

denoted in roman numerals
- refers to construction features

## Code Development

often created as a reaction to a disaster
- effort at this time to be proactive instead of reactive
  - addressing issues such as electric cars

committees form to address issues
- members author code language
- draft copy of code circulated
- challenges received and either
  - accepted
  - accepted with revision
  - rejected
- final draft presented when whole code body meets
  - hearings held
  - membership votes to accept or reject code provisions
    - item by item

*Note:* The following are some areas of law that are important to all firefighters.

## Operation of Emergency Vehicles

1988 federal legislation requires operators of vehicles over 26,001 pounds GVW or towing trailers over 10,000 pounds have a Class B license

state law
- when operated in non-emergency, fire vehicles must obey all traffic laws

certain exemptions when *authorized emergency vehicles* are en route to an emergency
- must be an authorized vehicle
- must display required lights and siren
- not exempt from exercising caution and due regard for safety
- within department policy
- to a true emergency

### Infectious Disease

Federal Rehabilitation Act of 1973
- prohibits discrimination against handicapped
- infectious diseases ruled a handicap in *Chalk v. U.S. District Court*
    - HIV and AIDS fall under this ruling
    - not allowed to disclose this information to anyone not directly involved in patient care
    - patients are not required to advise you of their condition
    - hospital is not allowed to advise you of their condition
    - cannot force patient to take a blood test

***wear your EMS PPE every time you deal with a patient, no exceptions!***

- trained personnel have a duty to act unless extreme hazard can be proven
    - HIV/AIDS is not considered to be an extreme hazard in most circumstances

### Good Samaritan Laws

protect persons acting within their training
- do not exceed your training
- be careful who you have assist you

### Personnel Safety

federal OSHA compliance instructions issued in May 1995
- SCBA required when performing interior structural fire fighting
- must operate in buddy system
    - in direct contact with each other while inside
- equipped and trained personnel must be available outside for rescue
- four personnel required at scene
    - two inside
    - two outside

### Scene Management

determined by law
- agency with primary investigative authority

traffic accident = police

fire = fire department

EMS may be given to fire department even if in police jurisdiction
   due to medical training

hazmat may be given to fire department even if in police jurisdiction
   due to hazmat training

## Review Questions *(answers appear in bold italics)*

1. The supreme law of the United States with which no other laws must conflict is the: ***United States Constitution.***

2. May a state law be different than a federal law? ***Yes, it may be more restrictive, but not less restrictive.***

3. If a postal vehicle, fire engine, and private auto arrived at an intersection at the same time, who would have the right of way? ***Use the relationship of the levels of law as your guide. The postal vehicle has the right of way.***

4. A failure to act is considered which type of feasance? ***Non feasance.***

5. What court case determines the responsibility of a fire prevention bureau to enter premises? ***See vs. City of Seattle***

6. How are model codes developed? ***Through a process of proposal and challenge.***

7. Who may suggest changes to model codes? ***Anyone.***

8. How are model codes adopted at the local level? ***By ordinance.***

9. What type of occupancy is a school? ***E, educational.***

10. What is meant by the term one-hour construction? ***The assembly can be expected to resist breakthrough by fire for a period of one hour.***

11. Why is the location of a building in relation to the property line important from the standpoint of stopping fire spread? ***The proximity of exposures to the fire building.***

12. What is meant by a Type A occupancy? ***Assembly.***

13. What are your responsibilities when responding to an emergency with red lights and siren activated? ***To drive with due regard for the safety of all persons and property.***

14. Is it legal to refuse treatment to a person with HIV who is bleeding when you are not equipped with full medical PPE? Justify your answer. ***Only if the contact can be proven to present an undue risk.***

15. If you act within the scope of your training at an accident scene while you are off duty and the patient dies, can you be held liable? ***Not if your state has a "Good Samaritan Law" in force.***

16. Under federal OSHA standards how many persons must be at the scene before interior fire fighting can be considered? What factors affect this decision? ***Four. The determining factor is if you are entering an atmosphere that is immediately dangerous to life and health.***

17. At the scene of a motor vehicle accident, who usually has scene management authority and responsibility? ***The agency with primary investigative authority.***

# Chapter 12

# Fire Protection Systems and Equipment

## Learning Objectives

*Upon completion of this chapter, you should be able to:*

- Describe the components of a water supply system.
- Explain the importance of a dependable water supply system.
- Describe the components and importance of a fire department water supply program.
- Describe fire detection systems and their components.
- Describe different types of extinguishing systems and their components.
- Describe the different types of extinguishing agents.
- Explain how the various types of extinguishment agents work to extinguish fires.

## INTRODUCTION

Water is the most common extinguishing agent used for combating fires. Numerous systems have been developed to effectively apply it to fires. Additives to make water an even more effective agent are utilized in many of the application methods. Other extinguishing agents and systems to apply them have been developed as well. An understanding of the agents and systems is necessary to effectively utilize them in fire control.

## LESSON PLAN

### Public Water Companies

> one of the most important single factors in municipal fire protection is water supply
>> water companies formed to ensure adequate community water supply
>> many are publicly held and act as monopolies
>>> only one water company in an area
>> fire or building department determines required fire flow for buildings
>> small systems may have a tough time handling large fire flow demand
>> requires close cooperation between F.D. and water company
>>> to boost pressure
>> need to notify F.D. when repairs or shut downs take place

### Private Water Companies

> usually in industrial or commercial situations
> maintains own distribution and storage

### Water Supply Systems

> must have storage capability and redundant system
>> factors for determining required capacity
>>> frequency and duration of drought
>>> danger to system from natural disaster
>> supply systems
>> gravity system
>>> river or reservoir above city level
>> direct pumping system
>>> water pumped from river or reservoir
>> underground storage
>>> water pumped from aquifer (water table)
>> combination system
>>> gravity and pump supply
>>> water stored in tanks to cut pump running time
>>> elevated tanks maintain pressure on system
>> duplication of equipment

to prevent system failure
generators to back up electrical supply
underground electrical supply to prevent damage
adequacy criteria
average daily consumption
figured over last 12 months
maximum daily consumption
highest demand in 24-hour period over last 3 years
usually 1.5 times average daily consumption
peak hourly consumption
maximum used in any given hour of a day
usually 2 to 4 times normal hourly rate
minimum recognized system for fire protection
250 gallons per minute for 2 hours
250 x 120 = 30,000 gallons

## Distribution System

from storage through treatment system into distribution system
underground piping of various sizes (water mains)
largest size are called primary feeders
widely spaced, supply for smaller mains
best when laid in grid pattern
looped to prevent dead ends and pressure drops
secondary feeders
reinforce the grid
concentrate supply in high demand areas
distributors
serve individual hydrants and blocks of consumers
commonly used main sizes
6- to 16-inch diameter
doubling size quadruples water flow
formula: diameter of large pipe squared divided by diameter of small pipe squared
6-inch pipe used if it does not exceed 600 feet in length and is connected in a grid pattern
valves to be placed maximum of 800 feet apart
prevents shutting down long parts of main

## Fire Hydrants

two basic types
wet barrel
*Transparency 29
barrel contains water

may have more than one valve
common fitting size 2½, 4, and 4½-inch
equipped with pentagon nuts
dry barrel
barrel dry to prevent freeze damage
valve located underground
special hydrant types
airport taxi way
total hydrant underground
dry hydrant
*Transparency 30
drafting source from static water source
facilitates drafting operation
spacing specified by local ordinance
250 feet in mercantile and 500 feet in residential
ISO maximums 330 feet and 660 feet
hydrant maintenance program
inspect, clean, and lubricate all parts
clear weeds and other obstructions
flush when necessary
notify water company
open and close valves slowly

*If at any time a valve through which water is flowing is closed quickly, water hammer can occur. Open and close all valves slowly!*

hydrant testing
done on new systems to check flow rates
done on old systems to check condition of system
hydrant painting
for visibility
to identify flow rates
green top 1000 gpm or more
orange top 500 to 999 gpm
red top up to 500 gpm
black cap dead end main
white top for hydrant out of service

## Water Systems Program

used to promote cooperation between water company and F.D.
Letter of Working Agreement
written and signed by officials of water company and F.D.
agreement for maintaining and testing system
F.D. performs painting and minor repairs of hydrants

water company performs major repairs
water map with description of system and its capabilities
showing main size and location
location of pumps, tanks, hydrants, and valves
hydrant records
flow, outlet size and service date

## Auxiliary Sources of Water Supply

lack of adequate water system
water available from static sources
cisterns, reservoirs, canals, tanks, swimming pools, and so on

## Private Fire Protection Systems

designed to protect individual properties from fire
Detection Devices
two main purposes
alert building occupants and notify fire department
alarms sent to F.D. through
direct tie-ins, security guards, monitoring companies
smoke detectors (alarms)

*Note:* A working smoke detector in the home doubles a person's chance of surviving a fire. Nearly one-half of the fires and three-fifths of the residential fire deaths occur in homes with no detector. *Source:* Facts About Fire in the U.S.

ionization chamber
ionizes air entering chamber
fire ionizes smoke causing alarm to activate

*Note:* Underwriters Laboratories recommends that battery-operated smoke alarms be replaced every five years and electrical system-operated be replaced every ten years. It is also recommended that batteries be replaced every six months. "Change your clock, change your battery" program.

flame or light detectors
ultraviolet measures light waves
infrared measures either flame flicker or infrared component of flame
visible smoke detectors
light beam interruption triggers alarm
rate of rise detector
when rise in temperature exceeds prescribed rate, alarm is triggered
fixed temperature detector
melts at preset temperature
manual pull station
water flow or excess flow switch
triggered when sprinkler heads flow water
combination of the above used in many occupancies

### Extinguishing Agents

### Water

most commonly used agent
- extinguishment by cooling
  - high specific heat
    - one gallon absorbs 1,280 BTU from 62° to 212° F
    - additional 8,080 BTU to convert water at 212° to steam at 212° F
    - most efficient when water is applied in fine droplets (fog stream)
  - extinguishment by smothering
    - volume expands 1,700 times when turned to steam

### Foam

concentrate added to water
- extinguishes three ways
  - cools surface
  - cuts off vapor production
  - insulates surface from radiated heat
- types of foam
  - chemical
    - combination of two powders, not commonly used anymore
  - mechanical
  - *Transparency 31
    - concentrate added to hose stream through inductor
    - finished product is foam solution
  - air is entrained to make bubbles in foam
    - special aeration devices
    - foam nozzles
    - air injection
- types of mechanical foam for Class B fires
  - AFFF causes water to float on liquids with lower specific gravity
    - 3% concentration for flammable liquids
    - layer one molecule thick
    - self healing

***never walk through the spill area even with a foam blanket in place!***

- AFFFATC for use on polar solvents
  - 3% for flammable liquids, 6% for polar solvents
  - would draw water from AFFF
- protein foam
  - animal matter
- fluorinated protein foam
  - fuel shedding characteristic

good for subsurface injection
combustible gas indicators should be used to ensure vapor suppression
high expansion foam
- made by running solution over screen and blowing air through it
- can obscure holes and objects
- do not get flammable vapors in fan

foam for Class A fires
- different than Class B
- not interchangeable
- concentration of 0.1% to 1%

wetting agents
- reduce surface tension of water
- help it to soak in to fuel
  - great for overhaul

## Fire Retardant

short term
- depends on wetness to suppress fire
  - class A foam
  - wetting agents
  - applied by air or ground-based units

long term
- depends on chemical reaction with fuel to suppress fire
- contain pigment to remain visible from air
- applied by air

carbon dioxide
- extinguishes by smothering
- small cooling effect
- works best in closed areas
- used where water or dry chemical can cause major damage

halogenated agents
- group 7 elements
- break chemical chain reaction
- trade name Halon followed by appropriate number to identify compound present
  - 1211, 1301

dry powder
- used on flammable metals
- may be **dry** sand in a pail, with a shovel to apply

***water is not commonly used on combustible metals as it may react violently!***

### Extinguishing Systems

#### Sprinkler Systems

    automatic systems have been proven 96% effective

    failures due to improper maintenance, inadequate or shut off water supply, incorrect installation or design and obstructions

        attack fire in incipient phase

    residential sprinklers

        required by ordinance in some jurisdictions

        less complicated and expensive than commercial systems

*Note:* The conflict over whether to require residential sprinklers continues in communities across the U.S. On one side, the fire department is usually for them; on the other side, building industry associations are against them because they add to the cost of construction of a home.

    commercial and industrial systems

        systems consist of several basic components

            open screw and yoke (OS&Y) valve

            post indicator (PI) valve

            main control valve

wet pipe system

    water behind sprinkler heads at all times

    clapper in control valve keeps water from reentering domestic supply

    alarm valve actuates when flow is detected

    retard chamber prevents false alarms due to pressure surge

dry pipe system

    used where freezing temperatures are a problem for wet systems

    compressed air in system keeps water below clapper valve in main valve body

    open head releases air and system becomes wet

        may be up to one minute delay in water discharge

deluge system

    all heads are open at all times

    sensor system releases water into system

    used where flooding amounts of water are required

        lumber operations, LPG storage, and so on

Fire department connection

*Transparency 32

    attached to riser

    for attaching fire department pumper

        to boost pressure or add volume to system

path of water through system

    water main

    post indicator and/or OS&Y valve

    main valve mounted on riser

    *Transparency 33

feed mains
cross mains
branch lines
sprinkler heads discharge onto fire

sprinkler heads
include a deflector to divide spray into droplets
various types for different applications
pendant works in hanging down position
upright works with head upright
sidewall works with head horizontal to pipe
not interchangeable in their installation

standpipe systems
usually found in stairwells
prevent having to lay hose up stairwells to fire floor
Class I—2½-inch hose connection for fire department use
Class II—1½-inch hose connection with hose for occupant use
Class III—2½ and 1½-inch connections for occupant and fire department use

foam systems
stationary or vehicle mounted
installed in aircraft hangars and other occupancies as fixed systems
create foam in one of four ways
*Transparency 31
1. inductor: drafts concentrate from container using venturi principle
2. injection: injects concentrate into water before or after pump
3. batch mixing: concentrate dumped into water tank
4. premixing: concentrate and water stored as a mixture
compressed air foam system (CAFS)
compressed air injected into hose stream after pump
anytime foam is used, great care must be taken to use the proper concentration and use the foam producing equipment within specification, or quality foam will not be produced

gas extinguishing systems

carbon dioxide systems
product stored in large cylinders or tanks
plumbed to nozzles
should have alarm system to evacuate area prior to discharge
follow up with water in Class A materials is often necessary

halogenated agents
installed much like $CO_2$ systems

dry chemical systems
powder stored in a container
stored pressure system has pressure in same container
remote reservoir system has expellant gas in attached storage container

can handle rough service and require little maintenance

used in race cars, on equipment, range hoods, home extinguishers, and so on

fire extinguishers

from small hand held to large wheeled types

contain all of the extinguishing agents listed above

fire pumps

used to boost pressure or ensure supply to system

centrifugal and vertical turbine pumps used

pressure reducing devices

used to reduce pressure in lower floors of high-rise buildings or systems with excess pressure

may require use of special nozzles for fire fighting

*always preplan for fire fighting needs with the systems installed in your area; your life may depend on it!*

## Review Questions (answers appear in bold italics)

1. What is the most common extinguishing agent used by firefighters? ***Water.***

2. What is the difference between a public and private water company? ***A public water company is owned by a public entity and private water companies are owned privately.***

3. When does the fire department get involved in planning fire flow for new construction? ***Usually before the buildings are built.***

4. List three components of a water supply system.
   1. ***Storage***
   2. ***Pressure system***
   3. ***Distribution***

5. What are the differences between direct pumping and gravity fed water systems? ***Pumped systems pull their water from the aquifer. Gravity fed systems take their water from storage sources that are located at a higher elevation.***

6. Why is duplication of water supply equipment necessary? ***When a pump goes down or a main is ruptured, the required water flow must still be available for normal and emergency use.***

7. List the names of the mains in a distribution system and show their relationship to each other. ***Primary feeders, secondary feeders and distributors. See diagram in text for relationship.***

8. Why is it necessary for a supply system to be gridded? ***For two reasons: the flow should still be available to most of the system if there is a ruptured main, and a gridded system allows the hydrants to be fed from two directions so two pumpers can operate off the same main.***

9. What is the difference between a wet barrel and dry barrel hydrant? ***A wet barrel hydrant contains water in the hydrant; a dry barrel hydrant has a valve that holds the water to below ground level and is equipped with a drain to prevent damage from freezing.***

10. What is a dry hydrant? ***It is a drafting connection at a static water source that is below the location of the connection.***

11. List the steps to be performed in hydrant maintenance.
    1. ***Remove caps and clean and lubricate threads.***
    2. ***Operate the hydrant to flush out any debris.***
    3. ***Lubricate the stem.***
    4. ***Turn off flow and replace caps.***

5. *Paint as necessary.*

12. How are hydrants flow tested? *By testing their residual pressure. A pitot gauge is used to measure the flow pressure and tables are used to determine the flow.*

13. A hydrant with orange painted caps should flow how much water? *Over 500 gpm, but less than 1,000 gpm.*

14. What are the component parts of a water systems program? *Water company agreements, system maps, and hydrant records.*

15. List two types of detection systems and explain their operation.
    1. *Ionization chamber*
    2. *Flame or light detector*
    3. *Visible smoke detector*
    4. *Rate of rise detector*
    5. *Fixed temperature detector*
    6. *Water flow/excess flow switch*
    
    See text for descriptions.

16. What are the differences between dry pipe and wet pipe sprinkler systems? *Dry pipe systems have air in the piping that holds back the water from entering. They are used in areas where freezing temperatures can negatively affect the system. Wet pipe systems have water in the piping.*

17. List two fire suppression systems that do not use water and explain how they work.
    1. *$CO_2$*
    2. *Dry chemical*
    3. *Wet chemical*
    4. *Halogenated agent*
    
    See text for descriptions

18. In terms of wildland use, list short-term and long-term retardants and explain how they are used in fire fighting.
    1. *Short-term retardants are water, wetting agents, and foam. These agents are applied directly on the flaming front and require immediate follow up to ensure extinguishment.*
    2. *Long-term retardants are applied from aircraft and work well when dry. They are applied well in front of the fire, but require follow up to prevent burn through.*

19. What is a foam inductor and how does it work? *A foam inductor introduces foam concentrate into the attached hose stream at a specified rate. The concentrate is picked up by venturi action in the inductor.*

20. What is the difference between AFFFATC and AFFF? *AFFFATC contains agents that prevent it from mixing with polar solvents, such as alcohols. AFFF is not effective on polar substances.*

21. Why is it important to know manufacturer's instructions when using foam concentrates and devices? *The purpose of using these items is to produce good-quality foam that will have the desired extinguishing effect. If the wrong type of concentrate is used, or the equipment is used incorrectly, quality foam will not be produced.*

# Emergency Incident Management

## Learning Objectives

*Upon completion of this chapter, you should be able to:*

- Explain the need for a plan at every incident.
- Differentiate between offensive, defensive and combination modes of attack.
- Explain the need for organized thought processes in incident assessment.
- Describe the strategic priorities at an incident.
- Explain the terms strategy, tactics, and tasks.
- Explain the need for size up of an incident.
- Explain how a size up is performed and what information it is necessary to communicate.
- Describe the NIIMS Incident Command System.
- Explain the need for unified command on a multijurisdictional incident.

## INTRODUCTION

Everyone assigned to an incident has an effect on the outcome. It is your responsibility to become trained in incident management techniques so that you can contribute in a positive way. Through the study and application of the information in this lesson, you can increase the incident effectiveness of yourself and your team. If you ever hope to be promoted to a management position in the emergency services, you need to start early in learning the concepts and use of incident management techniques.

## LESSON PLAN

### Management Responsibility

> first-in officer initiates the plan
>
> all at scene must be alert and aware of the plan and the hazards present

*fire trucks are not used to deliver victims to the scene!*

> everyone is required to participate in the decision-making process

### Incident Planning

> every incident *must* have a plan
>> to achieve effective utilization of resources
>> to resolve the incident with as little damage as possible
>
> first decision is to establish objectives
>> depends on incident type
>
> next step is to determine the strategies to accomplish the objectives
>
> above all, plans must be flexible to address changes in the incident as it progresses
>
> attack modes
>
> offensive: aggressive, direct attack
>> interior attack on structure, direct attack on wildland
>
> defensive
>> protecting exposures, indirect attack on wildland
>
> combination
>> may be applied on different parts of the fire
>>
>> must be carefully coordinated to avoid conflicting actions
>>
>> requires clear communication
>>> between command and forces and among forces
>
> modes may have to be shifted during attack to account for changes in incident
>> possible structural collapse, wind shifts, changes in fuel, fire going out
>
> 7 strategic priorities
>
> identified by Lloyd Layman in his book *Fire Fighting Tactics*
>
> RECEO SV
>> 1. Rescue is always the first priority
>> 2. Exposures
>> 3. Confinement

4. Extinguishment
5. Overhaul
* Salvage
* Ventilation

salvage and ventilation are not numbered because they are started when necessary to accomplish the other priorities

tactics

to accomplish the objectives, tactics are implemented

examples:
    for rescue, hand lines are advanced in an interior attack
    for salvage, covers are spread
    for confinement, ventilation above the fire is performed

tasks

tasks are the individual jobs performed
    advancing a hand line is a task
    felling a snag is a task

*Note:* Use an example of an attack to illustrate relationship of objectives, strategies, tactics and tasks. Emphasize how this requires the use of a plan that is communicated to all of the participating resources.

## Size Up

size up is an *ongoing* mental process that results in a plan

FPODP
1. Facts
2. Probabilities
3. Own situation
4. Decision
5. Plan of operation

once the plan is in operation, the size up continues
    more facts become known
    probabilities and own situation change
    decisions are made and plan of operation must change as incident dictates

one should always critique incidents to learn what went right as well as what went wrong

don't make the same mistake twice!

## Vegetation Fire Size Up

1. correct location
2. size
3. fuel type
4. slope and aspect
5. rate of spread

6. exposures
7. weather conditions
8. potential of the fire
9. additional resource needs
10. objectives

### Structure Fire Size Up

1. correct location
2. height/stories
3. size
4. type of structure
5. location and area involved
6. level of involvement (percentage)
7. exposures
8. potential of the fire
9. additional resource needs
10. objectives
11. obtain an "all clear"

### Incident Command System

National Interagency Incident Management System (NIIMS)
- National Interagency Fire Qualification System (NIFQS)
  - qualifications, training and certification of personnel
- Incident Command System (ICS)
  - basic organizational structure for all types of emergencies
    - incidents may be large or small in size
    - simple or complex
- components of the ICS
  - common terminology
    - organizational functions
    - resource elements
    - facilities
  - modular organization
    - *Figures 13-1, 13-2 illustrate operations organization, top to bottom
      - expands or contracts depending on incident needs
  - integrated communications
    - managed through communications plan
  - unified command structure
    - regardless of jurisdictions involved
      - single agency/single jurisdiction
      - multi-agency/single jurisdiction

multi-agency/multi-jurisdiction
consolidated action plans
- all resources work for the operations section chief
- written plan to facilitate communication of objectives

manageable span of control
- based on safety and management issues

predesignated incident facilities
- Incident Command Post (ICP)
  - coordination of incident activities
  - only one on each incident
- Staging Area (staging)
  - resources available to respond in 3 minutes or less
- Incident Base
  - incident support activities (food, fuel and supplies)
- Camp
  - established on long duration incidents
  - a large forest fire may take more than a week to control

comprehensive resource management
- single resource
- task force: resources temporarily assembled for a specific mission
- strike team: set number of resources of the same kind and type with an established minimum number of personnel (refer to Appendix E)

resource status

assigned: in use at the incident

available: able to respond in three minutes or less (staged)

out-of-service: not ready for available or assigned status. Does not mean broken down. May still be called up to available status at any time.

## Organization

five areas *Transparency 34

1. Command: responsible for overall incident management
   - incident commander and command staff
   - safety officer
   - liaison officer
   - public information officer

The next four Chiefs (2–5) are the General Staff

2. Operations Section *Transparency 35
   - operations chief and subordinates
   - staging area manager: where "available" resources are kept
   - branch director: in charge of divisions and/or groups

  division: based on geography
    one side of a building
    one floor of a high-rise
    a length of fire line (from one point to another)
  group: functional in nature, may cross divisional boundaries
    ventilation group
    rescue group
    firing group
3. Planning Section *Transparency 36
  plans chief and staff
  resources unit: keeps track of resources at incident
  situation unit: current and predicted status of incident
  documentation unit: all documentation of incident
  demobilization unit: prepares demob plan and implements
  technical specialists: as required by incident
    structural engineers
    hazardous materials specialists
    fire behavior specialists
    weather specialists
4. Logistics Section *Transparency 37
  logistics chief and staff
    Service branch
      communications unit: communications plan
      medical unit: first aid for personnel
      food unit: provides meals and drinking water
    Supply branch
      supply unit: incident supplies
      facilities unit: incident facilities
      ground support: fuel and ground transportation
5. Finance Section *Transparency 38
  finance chief and staff
    time unit: personnel and equipment timekeeping
    procurement unit: purchasing
    compensation/claims unit: settles claims for damage
    cost unit: cost analysis data
advantage of this system
  personnel needed are trained and identified before the incident occurs
three questions you must be able to answer at any time at an incident
  What do you have?
  What do you need?
  What is your plan?

## Review Questions  *(answers appear in bold italics)*

1. What position is in charge at an incident? ***The Incident Commander (IC).***
2. Define objectives. ***Something towards which effort is directed.***
3. Define strategy. ***The art of devising plans to achieve a goal or objective.***
4. Define tactics. ***The art of employing and directing resources to achieve the objectives.***
5. What are the three attack modes?
   1. ***Offensive***
   2. ***Defensive***
   3. ***Combination***
6. Interior attack is performed in which mode? ***Offensive.***
7. List the seven strategic priorities.
   1. ***Rescue***
   2. ***Exposures***
   3. ***Confinement***
   4. ***Extinguishment***
   5. ***Overhaul***
   6. ***Salvage***
   7. ***Ventilation***
8. List the steps in a size up. Give a listing of the information necessary for structure and wildland size ups.
   1. ***Facts***
   2. ***Probabilities***
   3. ***Own situation***
   4. ***Decision***
   5. ***Plan of operation***

   ***Structure size up: correct location, height/stories, size, type of structure, location, and area involved/level of involvement, exposures, potential of fire, additional resource needs, and objectives.***

   ***Wildland size up: correct location, size, fuel type, slope and aspect, rate of spread, exposures, weather conditions, potential of the fire, additional resource needs, and objectives.***

9. What are the two main components of the NIIMS?
   1. ***National Incident Fire Qualification System (NIFQS)***
   2. ***Incident Command System (ICS)***
10. Whose responsibility is it to implement the ICS on incidents? ***The Incident Commander.***
11. Why is unified command important in the case of a multi-agency or multi-jurisdiction incident? ***To ensure that all available resources are applied in a coordinated attack.***
12. What are the positions of the command staff?
    1. ***Incident Commander***
    2. ***Safety Officer***
    3. ***Information Officer***
    4. ***Liaison Officer***
13. What are the positions of the general staff?
    1. ***Incident Commander***
    2. ***Operations Chief***
    3. ***Plans Chief***
    4. ***Logistics Chief***
    5. ***Finance Chief***
14. What is the difference between a group and a division? ***Groups are functional, divisions are geographic.***
15. What is the difference between a strike team and a task force? ***A strike team is a set number of resources of the same type with common communications and a leader. Example: dozer strike team, engine strike team. A task force is composed of resources assembled for a specific task with common communications and a leader. Example: water tender and two engines.***
16. What does it mean to be in available status? ***Able to respond from staging in 3 minutes or less.***
17. When operating an engine on an ICS managed incident, which unit do you see for fuel? ***Ground Support Unit, part of the Support Branch of the Logistics Section.***

# Emergency Operations

## Learning Objectives

*Upon completion of this chapter, you should be able to:*

- Identify the role of the fire department at various types of emergencies.
- List limitations of the fire department in certain emergency types.
- List important safety considerations when operating at different types of emergencies.

## INTRODUCTION

One of the fundamental roles of the fire department is to handle emergencies. Not every call is an emergency, nor is it the responsibility of the fire department to handle every emergency. The purpose of this lesson is to look at different incident types and some of the safety considerations they demand.

## LESSON PLAN

### Personnel

> which personnel respond depends on emergency type
> variety of emergency responders
>> law enforcement, fire, EMS, hazmat
>
> F.D. may or may not be in charge

### Structure Fire Fighting

> basic F.D. responsibility
>
> most departments spend the bulk of time and money preparing for this one function
>
> many safety considerations about these types of responses
>> what are the contents?

*SCBA will not protect you from skin contact with harmful chemicals!*

> leather items absorb chemicals
> roof loads
> interior hazards

*one of the worst interior structural hazards is stairs!*

> suspended ceilings
> furniture
> pets
> electricity
> structural collapse
> backdraft and flash over

*as you walk around a structure fire, do not walk upright in front of windows!*

*always leave yourself a second way out!*

*do not freelance!*

> do not use elevators unless you are sure where they will open
> exterior hazards
>> falling glass
>> raising ladders—always look up first
>> power lines

### Electrical Installations

do not enter without being accompanied by power company personnel
carcinogens may be present

### Wildland Fire Fighting

backfiring: removing fuel between head of fire and control line by burning it
burning out: removing fuel between flanks and heel of fire and control lines by burning it
neither of these should be attempted by untrained personnel
The Ten Standard Fire Fighting Orders *Transparency 39

- **F**ight fire aggressively, but provide for safety first.
- **I**nitiate all action based on current and expected fire behavior.
- **R**ecognize current weather conditions and obtain forecasts.
- **E**nsure instructions are given and understood.
- **O**btain current information on fire status.
- **R**emain in communication with crew members, your supervisor, and adjoining forces.
- **D**etermine safety zones and escape routes.
- **E**stablish lookouts in potentially hazardous situations.
- **R**etain control of yourself and your crew at all times.
- **S**tay alert, keep calm, think clearly, and act decisively.

The 18 Situations That Shout "Watch Out" *Transparency 39

1. The fire is not scouted and sized up.
2. You are in country not seen in daylight.
3. Safety zones and escape routes are not identified.
4. You are unfamiliar with the weather and local factors influencing fire behavior.
5. You are uninformed on strategy, tactics, and hazards.
6. Instructions and assignments are not clear.
7. No communications link with crew members/supervisor.
8. Building line without safe anchor point.
9. Building fire line downhill with fire below.
10. Attempting frontal assault on fire.
11. Unburned fuel between you and the fire.
12. Cannot see main fire, not in contact with anyone who can.
13. On a hillside where rolling material can ignite fuel below.
14. Weather is getting hotter and drier.
15. Wind increases and/or changes direction.
16. Getting frequent spot fires across the line.
17. Terrain and fuels make escape to safety zones difficult.
18. You feel like taking a nap near the fire line.

TRIAGE *Transparency 41

Take time to evaluate water needs versus availability.

Recon safety zones and escape routes.

Is the structure defendable based on construction type, topography, and anticipated fire behavior?

Are flammable vegetation and debris cleared within a reasonable distance?

Give a fair evaluation of the values at stake versus resources available and do not waste time on the losers.

Evaluate the safety risk to the crew and the equipment.

PROTECTION *Transparency 42

Park engines backed in so a rapid exit can be made if necessary.

Remember to maintain communication with your crew and adjoining resources.

On occasions when you are overrun by fire, use apparatus or structures as a refuge.

Tank water should not get below 50 gallons in case it is needed for crew protection.

Engines should keep headlights on, windows closed, and outside speakers turned up.

Coil a charged 1½-inch hose line at the engine for protection of crew and equipment.

Try not to lay hose longer than 150 feet from your engine.

It is important to keep apparatus mobile for maximum effectiveness.

Only use water as needed and refrain from wetting ahead of the fire.

Never sacrifice crew safety to save property.

four most common causative factors in tragedy and near-miss wildland fires

1. usually happen on small fires or deceptively quiet sectors of large fires

2. happen in light fuels, such as grass or brush

3. unexpected shift in wind direction or speed

4. when fires run uphill

LCES *Transparency 43

Lookouts

Communications

Escape Routes

Safety Zones

three ways to look

Look up

Look down

Look around

### Oil Fire Fighting

primary objective to extinguish fire and control source of leaks

***firefighters should be accompanied by refinery employees anytime they enter a refinery for firefighting purposes!***

three main problems in crude oil tank fires
> boil over occurs when hot oil contacts subsurface water in tank
> slop over occurs when oil is forced over tank edge by direction of hose streams
> froth over occurs when hose streams stir up surface of hot oil

extinguishment methods
> subsurface injection of foam through a manifold
> direct application of foam to burning surface
> hose streams directed to cool exposure tanks
> floating roof tank seal fire may be extinguished with fire extinguisher

gasoline spills
> foam to seal vapors

***do not walk in the area of the spill; it may ignite as you disturb the foam blanket!***

LPG
> cylinders are found in many settings
> vapor is heavier than air
> when pressure increases, relief valve provided to prevent over pressure
> relief valve may not release over pressure
> BLEVE can occur

***container pieces have been known to fly as far as one-half mile in BLEVE's!***

> utilize un-manned monitors

natural gas
> flammable gas
> vapor lighter than air
> control ignition sources and dissipate vapors with water fog (same for LPG)

## Hazmat

approximately 2,000 new chemicals produced each year
hazardous materials may be present in almost any type of incident
every incident requires a precautionary approach
> upwind, uphill, and upstream
> safety must always be the first operational priority

federal law requires the establishment of an incident command system on any incident involving hazardous materials
> F.D.'s primary objectives are to isolate, identify, and deny entry
> back equipment in for quick exit
> set up perimeters
>> exclusionary (hot) zone
>> contamination reduction (warm) zone

support (cold) zone

passing from hot to warm zone requires decontamination

    F.D. personnel and victims

information on positively identified materials available from various sources

    CHEMTREC

    DOT Emergency Response Guidebook

*Note:* The new Guidebook, 1996 *North American Emergency Response* edition, is quite different than it used to be. Older versions of the guidebook should not be used.

    data bases and chemical dictionaries

    data is usually given for a single chemical, in a pure state, under laboratory conditions

    F.D. responds to accidental releases

    chemicals are often mixed with other materials when spilled

    may not react as predicted

    worst case scenario must always be considered

common F.D. actions at scene

    dike, divert

    control if material identified and suitable PPE is available

## EMS

in many departments this may be upward of 60–70% of calls

***one of the primary problems involved in providing emergency medical assistance is avoiding exposure to bloodborne and airborne pathogens!***

    proper PPE must be worn every time

    treat every victim as though he or she is infectious

    "If it's wet and it isn't yours, don't get it on you."

    turnouts are not the best protection

    PPE are required when working incidents where sharp edges and fire are a hazard

    never carry contamination back to the station or home with you

        properly dispose of contaminated materials

## Vehicle Accidents

    dangers from spilled fuel

        slipping

        fire

        park uphill and upstream of spill

        pull an attack line and charge it, keep it manned

    dangers from passing vehicles

        use fire engine as a shield between you and oncoming traffic

    dangers from power rescue tools

exert tremendous force (10,000) pounds

can send pieces of trim flying

when a piece of the vehicle finally breaks loose it can move with a lot of energy

should only be used by trained personnel

new items that pose a danger to firefighters

air bags

electric vehicles

## Vehicle Fires

approach from front or rear quarter

avoid exploding tires and 5 mph bumpers

full PPE, including SCBA, always

## Aircraft Fire Fighting

many on-board hazards in larger planes

oxygen systems

fuel

hydraulic fluid

magnesium wheels

military aircraft may carry ordnance (broken arrow)

F.D. operations

clear a path

make entry and effect rescue

extinguishment and overhaul

## EMS and Fire Fighting with Aircraft

helicopters are a common tool

require special precautions

*Transparency 44

1. Approach and depart helicopters from the side or front in a crouching position, in view of the pilot.
2. Approach and depart the helicopter from the downhill side to avoid the main rotor.
3. Approach and depart the helicopter in the pilot's field of vision, do not go anywhere near the tail rotor.
4. Use a chin strap or secure your head gear (hard hat) when working under the main rotor.
5. Carry tools horizontally, beneath waist level to avoid contact with the main rotor.
6. Fasten your seat belt when you enter the helicopter and refasten it when you leave. A seat belt dangling out of the door can cause major damage to the thin aluminum skin of a helicopter.
7. Use the door latches as instructed. Use caution around plexiglass, antennas, and any moving parts.

8. When entering or exiting the helicopter, step on the skid. If you place your foot next to the skid and the weight of the ship changes, it may run over your foot.

9. Any time you ride in a helicopter in a wildland fire situation you are required to wear full PPE.

10. Do not throw articles from the helicopter as they may end up in the main or tail rotors.

    setting up landing zone (LZ)

    dust abatement

    wind indicator

    advise pilot of hazards in area

    shut down traffic

***contact with any moving part of an aircraft is often fatal!***

wildland fire fighting with helicopters

    rotor down wash can fan flames

    can knock limbs from trees

    dropping water can knock you down

    exit the drop area whenever possible

    stay out from under helicopters and sling loads

wildland fire fighting with fixed wing aircraft

    3,000 gallons at 8 pounds a gallon dropped from 200 feet above ground at 130 mph

    wing tip vortices can fan flames and uproot or tear limbs from trees

    if you are going to get hit (get pink)

        lie face down

        head toward approaching aircraft

        lay tool aside

        hold hard hat on and cover head with both arms

        beware of slipperiness of retardant when you go back to work

        wash off vehicles as soon as possible

fire fighting with aircraft is expensive

    cost per hour for aircraft (1995) prepared by Cleveland National Forest, So. Calif.

    hourly flight rates and cost per gallon of water delivered

| | | |
|---|---:|---:|
| Sky Crane Helitanker | $7,365.00 | $0.64 |
| Bell 206 helicopter | 343.00 | 0.71 |
| DC4 airplane | 1,253.00 | 1.18 |
| S2 airplane | 700.00 | 1.26 |

    these figures illustrate why helicopters are gaining in popularity as fire fighting tools

## Review Questions  (answers appear in bold italics)

1. List at least three interior hazards encountered at structure fires.
   1. *Broken glass*
   2. *Smoke and heat*
   3. *Falling debris*
   4. *Structural collapse, etc.*

2. List at least three exterior hazards encountered at structure fires.
   1. *Holes*
   2. *Falling items*
   3. *Electrical wires*
   4. *Dogs*
   5. *Traffic, etc.*

3. What is wrong with freelancing at emergency scenes? *When your supervisor does not know where you are, he or she may not be able to find you when you get in trouble. All attacks must be made in a carefully coordinated fashion to prevent injuries.*

4. Why should you not enter electrical substations without electrical company personnel? *It is very easy to touch the wrong thing and get electrocuted.*

5. Using your text, list the 10 Standard Fire Fighting Orders. *See text.*

6. Using your text, list the 18 Situations That Shout Watch Out. *See text.*

7. What are the four components of LCES?
   1. *Lookouts*
   2. *Communications*
   3. *Escape routes*
   4. *Safety zones*

8. What are the three ways to "look"?
   1. *Look up*
   2. *Look down*
   3. *Look around*

9. List several of the safety considerations when providing structure protection. *See "PROTECTION" section in text.*

10. What are the three "overs" referred to in oil fire fighting?
    1. *Boil over*
    2. *Slop over*
    3. *Froth over*

11. What is the meaning of the term "BLEVE"? *Boiling Liquid Expanding Vapor Explosion.*

12. How should all suspected hazardous materials incidents be approached? *With a precautionary approach, from uphill, upwind, and upstream. Equipment should be backed in to allow rapid retreat.*

13. What is meant by "universal precautions" in regard to EMS incidents? *The wearing of full EMS PPE on every incident.*

14. Why should an attack line be pulled on all vehicle rescue situations? *In case there is a fire.*

15. What are the three priorities in an aircraft fire fighting incident?
    1. *Create a path*
    2. *Make entry and perform rescue*
    3. *Complete extinguishment and overhaul*

16. What action should you take if an air tanker is about to make a drop on your position? *Lie face down facing the approach direction on level ground, place your hand tool away from you, and cover your head with your hands.*

17. What are the safety rules for approaching a helicopter? *There are 10, see text.*

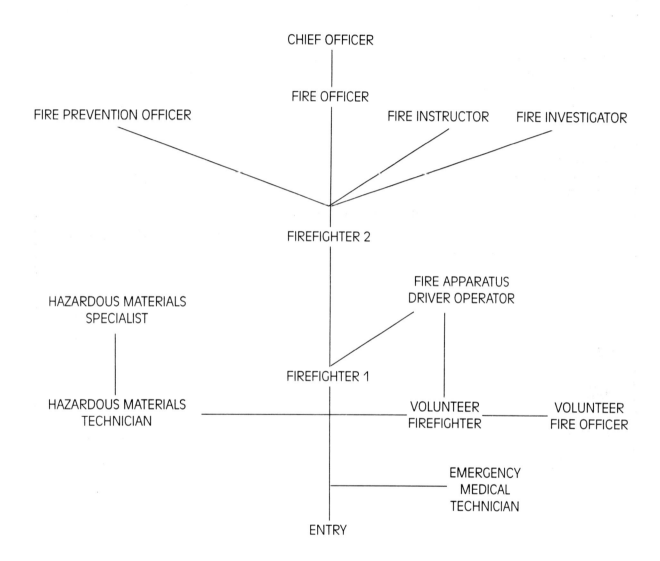

**Transparency #1** *Career ladder showing education certifications. (Taken from Figure 1-6 in the text)*

| NOTIFICATION OF RECRUITMENT | | Item No. _____ |
|---|---|---|

_____
(Print)   Last Name          First Name          Middle Initial

Title of Job _____

ARE YOU WILLING TO WORK SHIFTS   ___ Yes   ___ No

DO YOU SPEAK SPANISH FLUENTLY   ___ Yes   ___ No          BAKERSFIELD ONLY ☐

Telephone No. _____   Date _____

The KERN COUNTY Civil Service Commission is presently recruiting for the above named job. Applications are obtainable at the office of the PERSONNEL DEPARTMENT, 1115 Truxtun Avenue, BAKERSFIELD, and must be filed on or before _____.

If you fail to file an application for this job, it will be necessary to complete another of these forms in the event you wish to be notified in the future for this type of work.

Advise us of changes of address. Notification is a courtesy. The Personnel Dept. is not responsible for cards that aren't mailed or mailed and not delivered.

IF NO RECRUITMENT OCCURS FOR THIS JOB WITHIN TWO YEARS THIS CARD WILL BE DISCARDED.

**THIS IS NOT AN APPLICATION**

Personnel 580 1310 054 (Rev. 7/92)

---

Place
Postage
Here

PRINT

NAME _____

ADDRESS _____

CITY and STATE _____

ZIP CODE _____

**Transparency #2**   *Job interest card, front and back.* (Taken from Figure 1-7 in the text)

# FIREFIGHTER #1900

**HOW TO APPLY**: Applicants interested in participating in this examination must complete an official City of Bakersfield Application For Employment (no copies). Applications must be **received and stamped** in the Human Resources Office, City Hall, 1501 Truxtun Avenue, Bakersfield, CA 93301 during the filing period listed below.

**FILING PERIOD**:
| | |
|---|---|
| Tuesday, August 8, 1995 | 8:00am - 5:00pm |
| Wednesday, August 9, 1995 | 8:00am - 5:00pm |
| Thursday, August 10, 1995 | 8:00am - 5:00pm |

The Human Resources Office will **NOT** accept applications prior to or after the filing period. Applications which are postmarked or FAXed will not be accepted.

**NOTE: EMPLOYMENT APPLICATIONS MUST BE PROPERLY COMPLETED IN ACCORDANCE WITH INSTRUCTIONS ON FACE OF APPLICATION FORM. ALL PERTINENT INFORMATION NEEDED TO DETERMINE THAT THE APPLICANT MEETS THE MINIMUM QUALIFICATIONS MUST BE SHOWN ON THE APPLICATION; OTHERWISE THE APPLICATION WILL BE REJECTED. RESUMES WILL NOT BE ACCEPTED IN LIEU OF COMPLETED APPLICATION.**

**MINIMUM QUALIFICATIONS**:
- **AGE**: Must be 18 years of age at time of written exam.
- **EDUCATION**: Must possess a high school diploma or G.E.D.
- **LICENSE**: Valid driver's license required. Possession of a valid California driver's license at time of appointment is required.
- **VISION**: Visual acuity in each eye not less than 20/40 vision without correction and must have normal color vision.
- **PHYSICAL CONDITION**: Good physical condition. Weight must be in proportion to height.

**SALARY**: $2,902 - $3,537 per month

**EXAMINATIONS: ALL APPLICANTS WILL BE NOTIFIED BY MAIL OF DATE, TIME AND PLACE OF EXAMS.**

**Written Exam: (Pass/Fail)** The written exam is the first phase of the examination process and may measure comprehension of oral and written material, mathematical ability and mechanical aptitude. *Note: Only those applicants with the top 100 written scores who achieve a minimum score of 70% will be invited to the Oral Appraisal Interview.*

**Oral Appraisal Interview: (Weighted: 100%)** Appraisal will be made of applicant's personal qualifications, education/training, and experience. A minimum rating of 70% is required to qualify for the eligible list. To qualify for placement on the Civil Service Eligible List, applicants must pass both the Written Exam and the Oral Appraisal Interview.

**Physical Agility Exam: (Pass/Fail)** The Physical Agility Exam will be administered <u>AFTER</u> the Eligible List has been certified. The top 50 ranking candidates on the Eligible List will be invited to participate in the physical agility exam. If additional candidates from the eligible list are needed throughout the effective period of the list, eligibles will be notified to appear for the Physical Agility Exam. Failure to pass the Agility Exam will disqualify an applicant from further consideration.

**Background Investigation: (Pass/Fail)** Prior to appointment, applicants must successfully complete an investigation of their personal history and background to determine suitability for the position of Firefighter with the Bakersfield Fire Department.

**Non-Smoking Policy**: Newly hired employees must be non-smokers. Prospective employees will be required to sign an affidavit indicating that they have not smoked during the twelve (12) month period prior to hiring by the City. Further, they shall agree that they will not smoke, either on or off duty, during the term of their employment with the City. Violation of the non-smoking agreement shall result in disciplinary action and possible termination of employment.

Posted: July 17, 1995

AN EQUAL OPPORTUNITY/AFFIRMATIVE ACTION EMPLOYER
WOMEN, MINORITIES, AND INDIVIDUALS WITH DISABILITIES ARE ENCOURAGED TO APPLY

---

The provisions on this bulletin do not constitute a contract expressed or implied and any provisions contained in this bulletin may be modified or revoked without notice.

**Transparency #3** *Job announcement for firefighter position.* (Taken from Figure 1-8 in the text)

**Transparency #4**  *Fire marks.* (Taken from Figure 3-3 in the text)

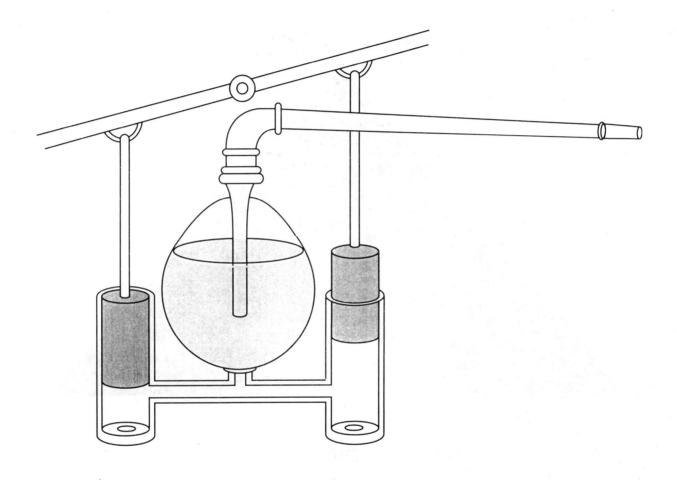

**Transparency #5** *Siphona using double acting piston pump.* (Taken from Figure 3-7 in the text)

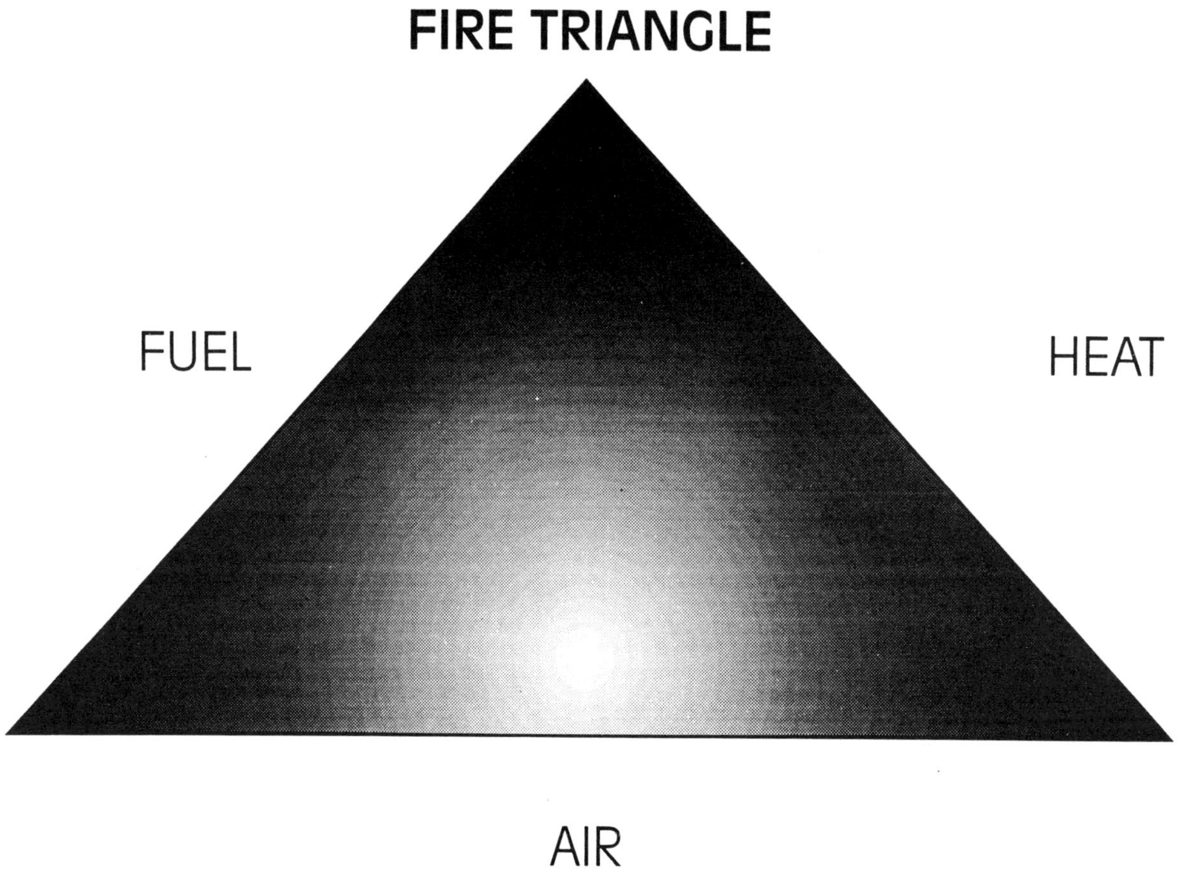

# FIRE TRIANGLE

FUEL

HEAT

AIR

**Transparency #6**  *Original fire triangle.* (Taken from Figure 4-1 in the text)

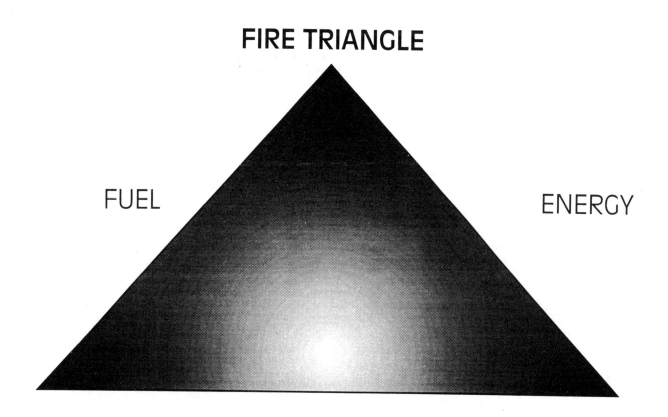

**Transparency #7** *New fire triangle.* (Taken from Figure 4-2 in the text)

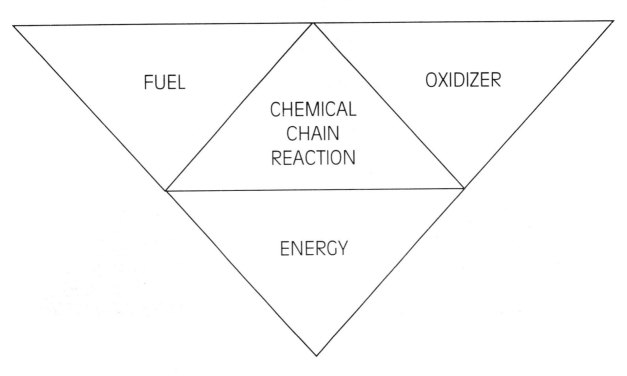

Transparency #8   *Fire tetrahedron.* (Taken from Figure 4-3 in the text)

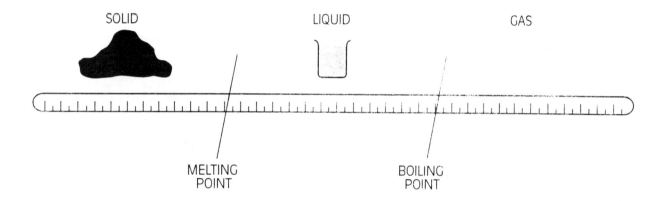

**Transparency #9**  *States of matter, in many cases, are temperature dependant.* (Taken from Figure 4-4 in the text)

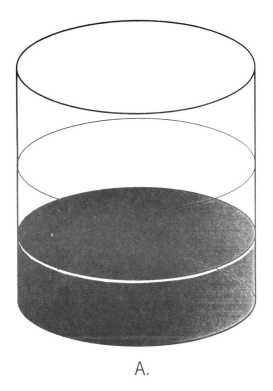
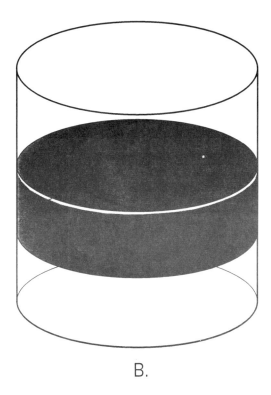

A.   B.

**Transparency #10**  *Specific gravity. Liquid in container A is heavier than water, liquid in container B is lighter than water.* (Taken from Figure 4-10 in the text)

Copyright © 1997 Delmar Publishers

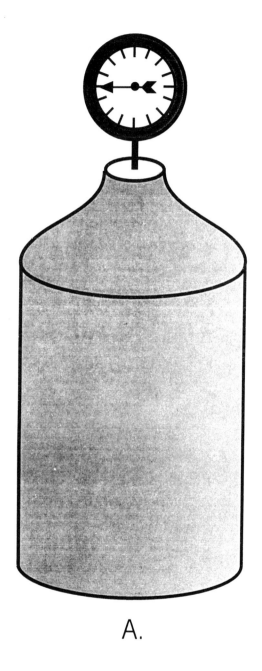 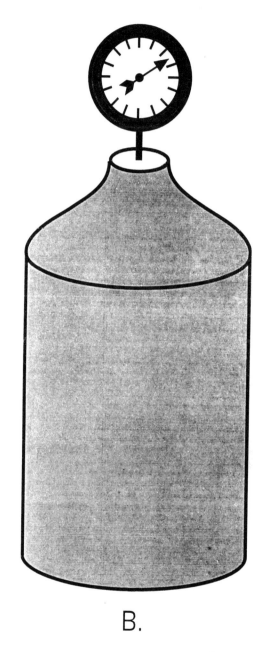

A.  B.

**Transparency #11** *Vapor pressure. Container A has less vapor pressure than container B* (Taken from Figure 4-11 in the text)

Copyright © 1997 Delmar Publishers

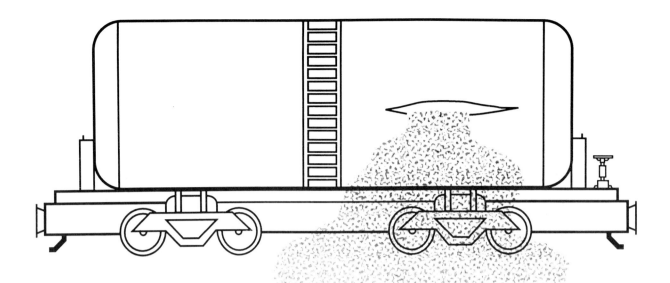

VAPOR DENSITY GREATER THAN 1

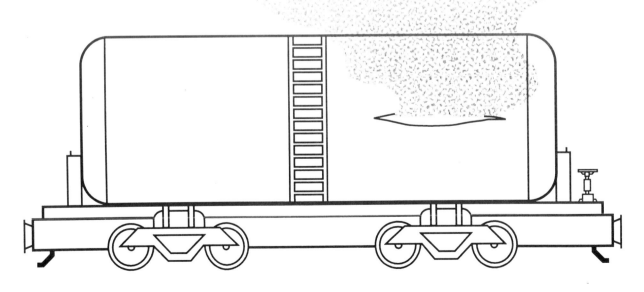

VAPOR DENSITY LESS THAN 1

**Transparency #12**  *Vapor density. Vapors lighter than air rise and those heavier than air sink.* (Taken from Figure 4-12 in the text)

Copyright © 1997 Delmar Publishers

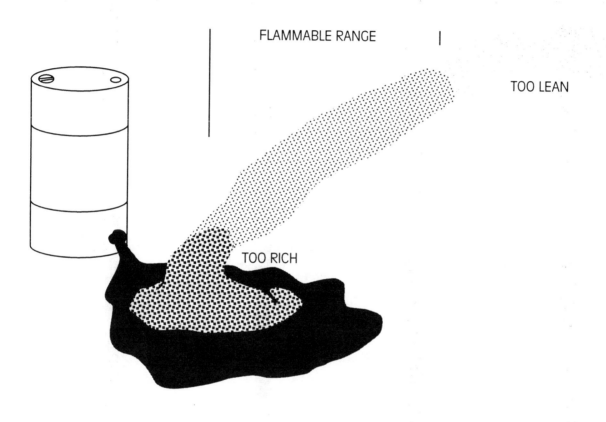

**Transparency #13** *Flammable range. Too rich to burn at liquid surface, within flammable range, and too lean to burn at increased distance from liquid surface. (Taken from Figure 4-13 in the text)*

Copyright © 1997 Delmar Publishers

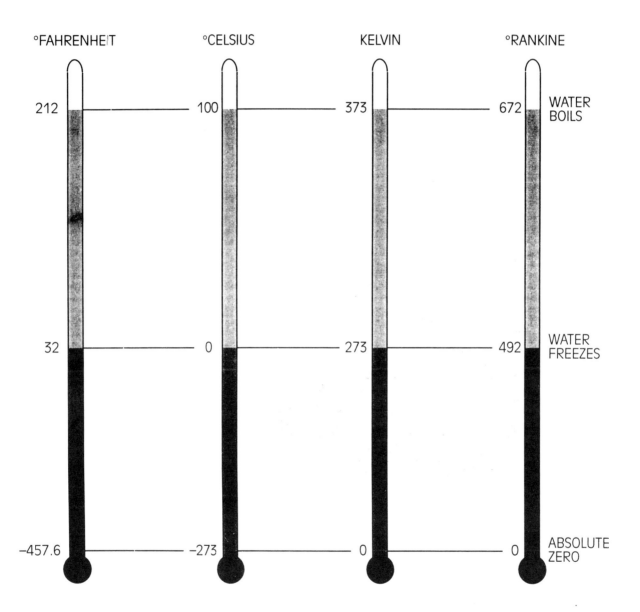

**Transparency #14** *Illustration of relativity of temperature measurement scales.* (Taken from Figure 4-14 in the text)

Copyright © 1997 Delmar Publishers

**Transparency #15** *Transfer of heat through conduction.* (Taken from Figure 4-15 in the text)

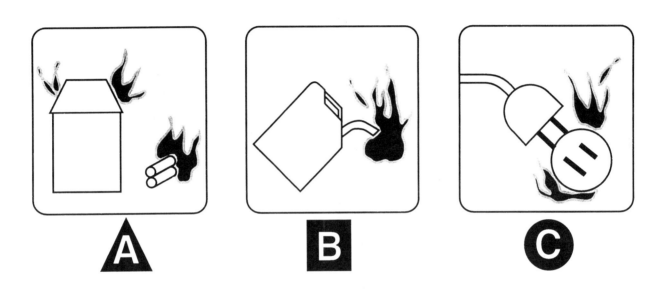

**Transparency #16** *Classification of fire symbols.* (Taken from Figure 4-19 in the text)

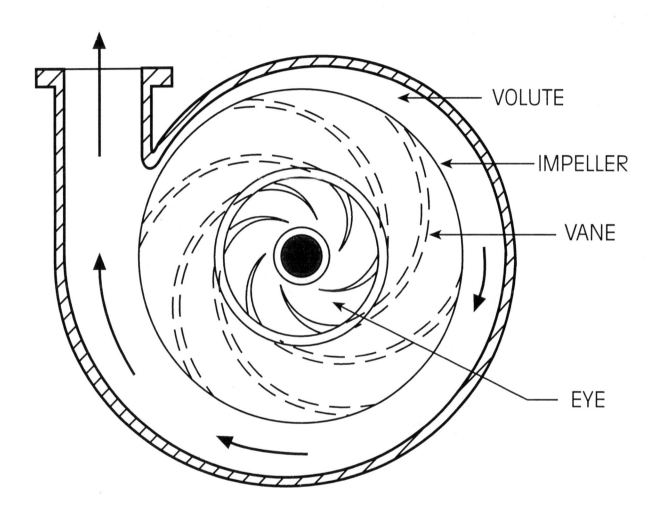

**Transparency #17** *Interior design of centrifugal pump.* (Taken from Figure 6-22 in the text)

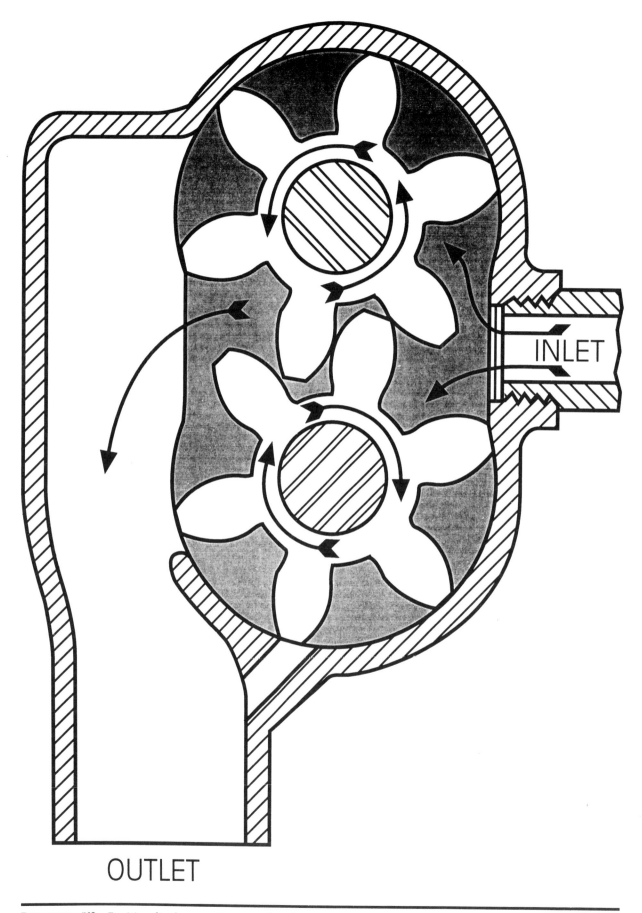

**Transparency #18** *Positive displacement pump—interior components.* (Taken from Figure 6-24 in the text)

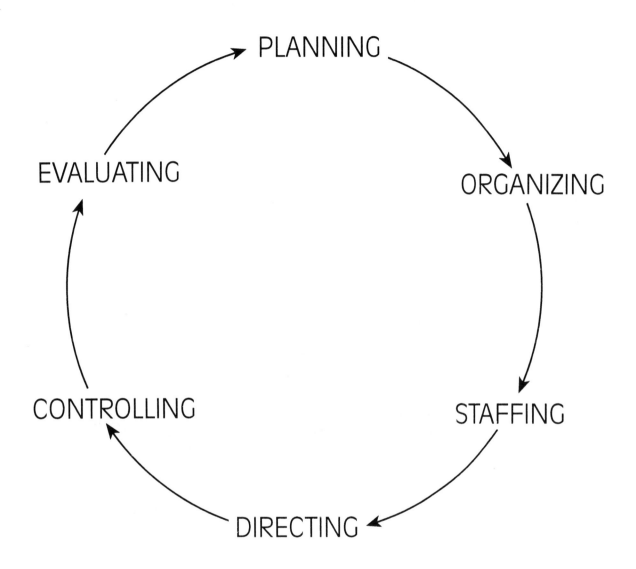

**Transparency #19** *The management cycle.* (Taken from Figure 7-2 in the text)

# STATION INSPECTION RECORD

STATION _____ SHIFT _____ DATE _____

CAPTAIN _____ BATTALION CHIEF _____ DIVISION CHIEF _____

✓ - O.K
X - SEE REMARKS

**1. PERSONNEL**
- A. Uniform
- B. Protective Clothing
- C. Grooming & Cleanliness
- D. Driver's License
- E. I.D. Card
- F. E.C. Card

**5. SAFETY**
- A. Overhead Storage
- B. Caution Signs
- C. Safety Bulletin Board
- D. Tool Storage
- E. Combustibles Storage
- F. Insecticide Chemical Stor.
- G. SCBA Tests
- H. Extinguisher(s)
- I. Smoke Detector(s)
- J. RADEF Equipment

**6. EXTERIOR**
- A. Paint
- B. Doors
- C. Sidewalks
- D. Ramps
- E. Parking Area
- F. Yard
- G. Hose Tower
- H. Gas Pump
- I. Flag Pole

**2. OFFICE**
- A. Files
- B. Maps
- C. Log Book
- D. Shift Change Log
- E. Desk
- F. Bulletin Board
- G. Tourist Information

**3. ENGINE HOUSE**
- A. Floor
- B. Work Bench
- C. Hose Rack
- D. Storage
- E. Turnouts
- F. Apparatus

**7. INTERIOR** (Office, Kitchen, Day Room, Dormitory, Rest Room, Heater Room)
- A. General Appearance
- B. Cleanliness
- C. Floors
- D. Windows
- E. Lights
- F. Furniture
- G. Storage Area(s)
- H. Woodwork
- I. Walls
- J. Ceiling

**4. PROGRAMS**
- A. Hazard Reduction
- B. In Service Inspections
- C. Pre Fire Plans
- D. Water Systems

REMARKS _____

Fire 580 2415 016 Catalog #9020 (Rev. 2/90)

**Transparency #20** *Station inspection form.* (Taken from Figure 7-8 in the text)

Copyright © 1997 Delmar Publishers

# EMPLOYEE PERFORMANCE REPORT
## PERSONNEL DEPARTMENT

| DEPT. NO. | DEPARTMENT NAME | CLASSIFICATION | EMPLOYEE NAME | EMPLOYEE NO. |
|---|---|---|---|---|
| | | | | |

REASON FOR RATING:

IF 6 MONTH PROBATIONARY DO YOU RECOMMEND PERMANENT APPOINTMENT? YES ☐ NO ☐ SPECIAL ☐ SEPARATION ☐

RATING PERIOD FROM _____ TO _____

## SECTION A - ITEMIZED CHECK LIST

EMPLOYEE'S IMMEDIATE SUPERVISOR SHOULD CHECK EACH ITEM IN THE APPROPRIATE COLUMN. REPORT MUST BE COMPLETED IN INK. ANY CHANGES MADE IN THE REPORT SUBSEQUENT TO THE EMPLOYEE'S SIGNING REQUIRE INITIALING BY THE EMPLOYEE AND PERSON MAKING THE CHANGES.

(READ INSTRUCTIONS ON BACK)

| | DOES NOT APPLY | OUTSTANDING | ABOVE STANDARD | STANDARD | IMPROVEMENT NEEDED | UNSATISFACTORY |
|---|---|---|---|---|---|---|
| **ALL EMPLOYEES:** | | | | | | |
| 1 ATTENDANCE | | | | | | |
| 2 PUNCTUALITY | | | | | | |
| 3 PHYSICAL FITNESS | | | | | | |
| 4 SAFETY PRACTICES | | | | | | |
| 5 PERSONAL NEATNESS | | | | | | |
| 6 COMPLIANCE WITH RULES AND REGULATIONS | | | | | | |
| 7 COOPERATION | | | | | | |
| 8 ACCEPTANCE OF NEW IDEAS AND PROCEDURES | | | | | | |
| 9 APPLICATION OF EFFORT | | | | | | |
| 10 INTEREST IN JOB | | | | | | |
| 11 ACCURACY OF WORK | | | | | | |
| 12 QUALITY OF JUDGMENT | | | | | | |
| 13 PUBLIC RELATIONS | | | | | | |
| 14 WRITTEN EXPRESSION | | | | | | |
| 15 ORAL EXPRESSION | | | | | | |
| 16 EQUIPMENT OPERATION | | | | | | |
| 17 NEATNESS OF WORK | | | | | | |
| 18 PERFORMANCE WITH MINIMUM SUPERVISION | | | | | | |
| 19 PROMPTNESS IN COMPLETING WORK | | | | | | |
| 20 VOLUME OF WORK PRODUCED | | | | | | |
| 21 PERFORMANCE UNDER PRESSURE | | | | | | |
| 22 PERFORMANCE IN NEW WORK SITUATIONS | | | | | | |
| **EMPLOYEES WHO SUPERVISE:** | | | | | | |
| 1 COORDINATING WORK WITH OTHERS | | | | | | |
| 2 ACCEPTANCE OF RESPONSIBILITY | | | | | | |
| 3 ESTABLISHMENT OF WORK STANDARDS | | | | | | |
| 4 TRAINING AND LEADING STAFF | | | | | | |
| 5 PLANNING AND ASSIGNING WORK | | | | | | |
| 6 FAIRNESS AND IMPARTIALITY TO STAFF | | | | | | |
| 7 CONTROL OF STAFF | | | | | | |
| 8 ADEQUACY OF INSTRUCTIONS | | | | | | |
| **ADDITIONAL ITEMS:** | | | | | | |

REVIEW INSTRUCTIONS ON REVERSE SIDE

## SECTION B - OVERALL PERFORMANCE

CHECK OVER-ALL EVALUATION WHICH MUST BE CONSISTENT WITH THE FACTOR RATINGS. ALTHOUGH THERE IS NO PRESCRIBED FORMULA FOR COMPUTING THE OVERALL PERFORMANCE, <u>SPECIFIC</u> WRITTEN COMMENTS ARE REQUIRED TO JUSTIFY <u>OUTSTANDING</u> OR <u>UNSATISFACTORY</u> RATINGS.

OUTSTANDING | ABOVE STANDARD | STANDARD | IMPROVEMENT NEEDED | UNSATISFACTORY

COMMENTS:

### EMPLOYEE'S CERTIFICATION: (CHECK ONE)

☐ I HEREBY CERTIFY I HAVE REVIEWED THIS REPORT. I UNDERSTAND MY SIGNATURE DOES NOT NECESSARILY MEAN I AGREE WITH ALL THE MARKINGS.

☐ I REQUEST AN APPOINTMENT TO DISCUSS THIS RATING WITH MY DEPARTMENT HEAD.

EMPLOYEE'S NAME _____ DATE _____

RATED BY _____ DATE _____

TITLE _____

REVIEWED BY _____ DATE _____

TITLE _____

PERSONNEL DEPARTMENT COPY

**Transparency #21** *Personnel evluation form.* (Taken from Figure 7-7 in the text)

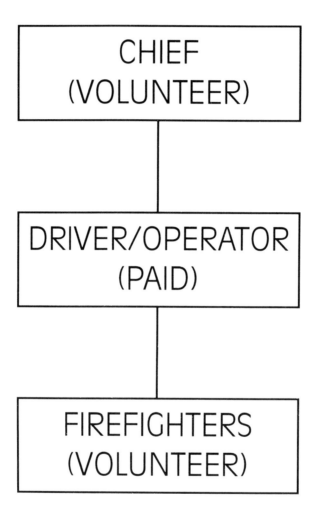

**Transparency #22** *Volunteer fire department organization with paid driver/operator position.* (Taken from Figure 7-3 in the text)

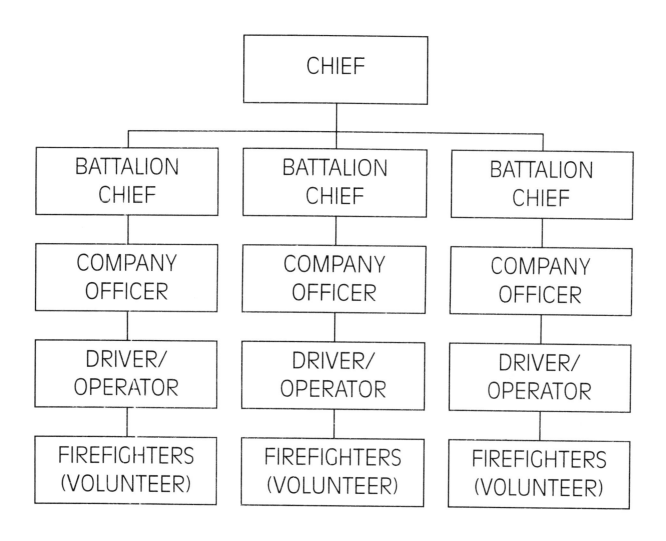

**Transparency #23** *Combination fire department organization.* (Taken from Figure 7-4 in the text)

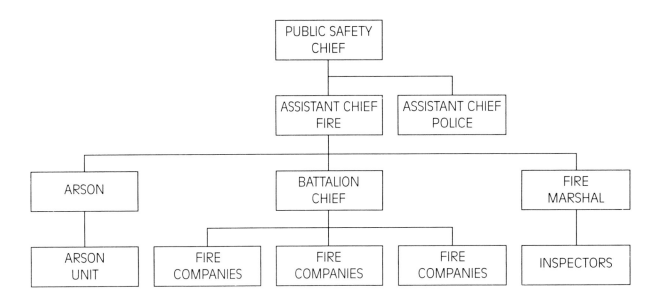

**Transparency #24** *Public safety department organization.* (Taken from Figure 7-5 in the text)

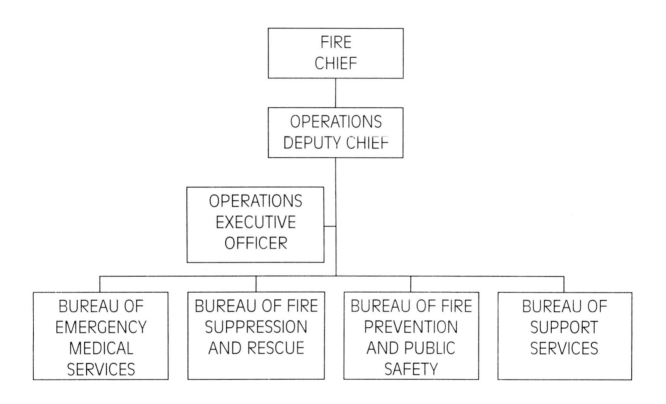

**Transparency #25** *Large paid fire department organization. Operations personnel are below the bureau of fire suppression and rescue chief.* (Taken from Figure 7-6 in the text)

# MONTHLY TRAINING RECORD
## FOR THE MONTH OF _____, 19____

BATTALION/STATION ____ SHIFT ____

SUBMITTED BY _____

1. CHIEF OFFICER
2. COMPANY OFFICER
3. ENGINEERS/TECHS
4. FIREFIGHTERS
5. TEAM BUILDING
6. EMS
7. SAFETY
8. READING
9. RESCUE
10. ARFF
11. VIDEO
12. HEALTH & FITNESS
13. ICS
14. FIRE INVESTIGATION
15. FIRE PREVENTION
16.* HAZ-MAT (PRO & TECH)
17. HAZ-MAT (SPECIALIST)
18. _____
19. _____
20. _____

| LAST, | FIRST | 1 | 2 | 3 | 4 | 5 | 6 | 7 | 8 | 9 | 10 | 11 | 12 | 13 | 14 | 15 | 16 | 17 | 18 | 19 | 20 |
|---|---|---|---|---|---|---|---|---|---|---|---|---|---|---|---|---|---|---|---|---|---|
| | | | | | | | | | | | | | | | | | | | | | |
| | | | | | | | | | | | | | | | | | | | | | |
| | | | | | | | | | | | | | | | | | | | | | |
| | | | | | | | | | | | | | | | | | | | | | |
| | | | | | | | | | | | | | | | | | | | | | |
| | | | | | | | | | | | | | | | | | | | | | |

**Transparency #26** *Training record.* (Taken from Figure 9-12 in the text)

# FIRE DEPARTMENT
## PREPLAN / INSPECTION REPORT

FILE NO.  MISC. NO.  New Inspection [X]  Purge [X]  H = HAZ/MAT  E = EXEMPT  D = DELETE  TARGET HAZARD  P = PRIMARY  S = SECONDARY  D = DELETE

GENERAL USE

D.B.A.

LOCATION  CITY  OCCUPANCY CLASS  PROPERTY CLASS

BUSINESS MANAGER  BUSINESS PHONE  NO. OF BLDGS.  NO. OF STORIES  BSMT. (B)  JURIS-DICTION

RESPONSIBLE SHIFT

BUSINESS OWNER  HOME PHONE  EMERGENCY PHONE  DATE OF LAST INSP.

Disposition Sym.   1 - Corrected   2 - Will Correct   3 - V.N. Issued. Call Back Necessary   4 - Citation Issued   5 - Referred

**BUILDING**                                   APPROVED   DISP.
                                               YES NA NO  SYM     REMARKS:
1. Exit Doors-Hardware Exit Signs-Lighted        1
2. Exit Corridors                                2
3. Aisle Spacing-Seating                         3
4. Occupant Load Signs                           4
5. Vertical Openings                             5

**COMMON HAZARDS**
6. Electrical                                    6
7. Furnace/Boiler Rooms-Heating Equipment        7
8. Cooking Equipment                             8
9. Decorations-Curtains-Drapes                   9
10. Housekeeping-Trash-Weeds                    10

**FIRE PROTECTION EQUIPMENT**
11. Fire Extinguishers                          11
12. Automatic Extinguishing Systems             12
13. Wet-Dry Standpipes                          13
14. Alarm Sysytem                               14
15. Fire Assemblies-Fire Walls                  15
16. Fire Sprinkler Records                      16

**OTHER HAZARDS**
17. Grease Hoods & Ducts                        17
18. Warning Signs                               18
19. Compressed Gas                              19
20. Other (see Remarks)                         20

**HAZARDOUS MATERIALS**
21. Training Records                            21
22. MSDS'S Available                            22
23. Chemical Inventory                          23
24. Labeling                                    24

REGULAR INSP. [ ]   TOTAL TIME TO COMPLETE INSPECTION _____

RE-INSPECTION [ ]   MISC. INSPECTION (PREVENTION ONLY)

MISC. INSP. [ ] (PREVENTION ONLY)

CLEARANCE
  GRANTED [ ]
  DENIED [ ]

SCHEDULED DATE OF RE-INSPECTION ___/___/___

OFFICE

OWNER / MANAGER
INSPECTOR   SHIFT   DATE

MONTH INSPECTION DUE

A = ANNUAL
O = ODD YEAR
E = EVEN YEAR

X - PREPLAN UPDATED [ ]

KCFD #97 CATALOG #9215

**Transparency #27**  *Fire prevention inspection form.* (Taken from Figure 10-11 in the text)

Copyright © 1997 Delmar Publishers

**Transparency #28** *Fire report form.* (Taken from Figure 10-12 in the text)

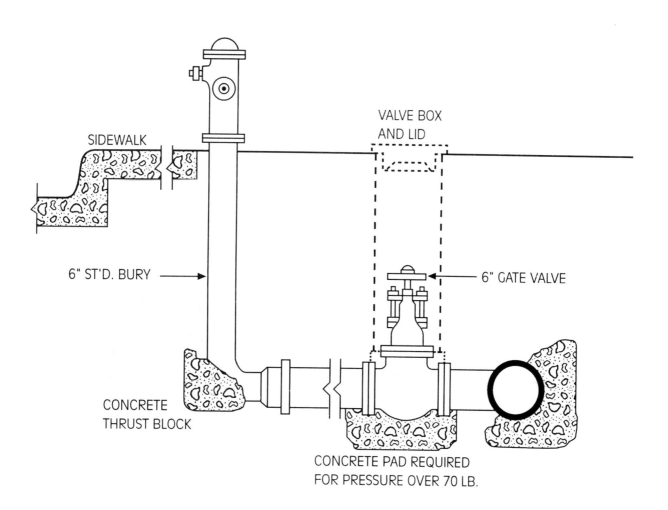

**Transparency #29** *Hydrant with underground plumbing.* (Taken from Figure 12-7 in the text)

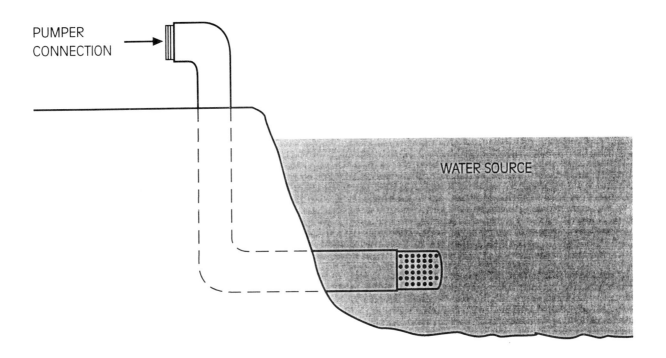

**Transparency #30** *Dry hydrant, suction source.* (Taken from Figure 12-6 in the text)

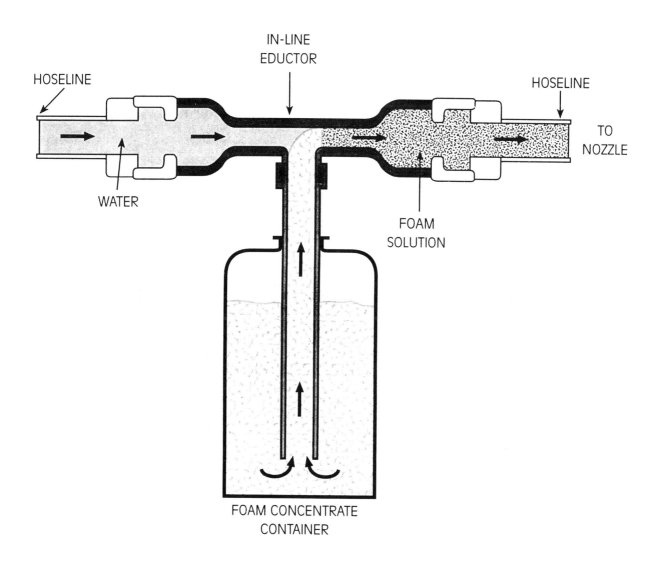

**Transparency #31** *Foam inductor.* (Taken from Figure 12-23 in the text)

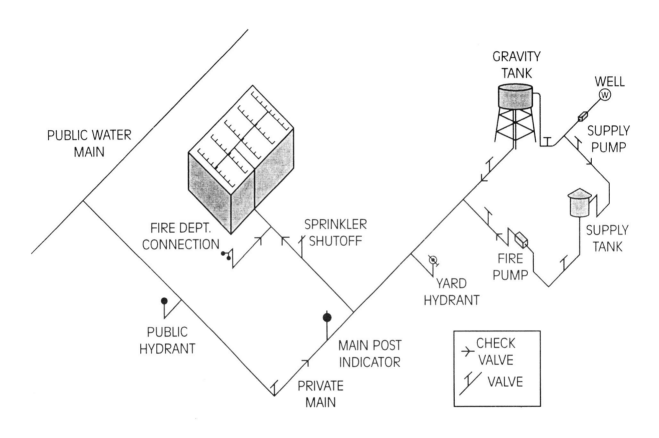

**Transparency #32** *Water system map.* (Taken from Figure 12-11 in the text)

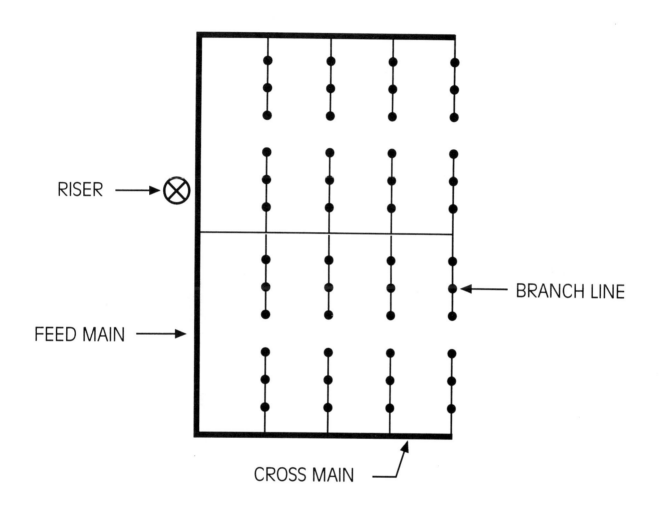

**Transparency #33** *Sprinkler piping diagram.* (Taken from Figure 12-20 in the text)

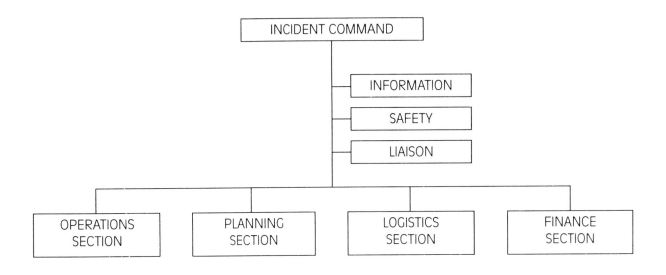

**Transparency #34**  *Incident command system command and general staff.* (Taken from Figure 13-1 in the text)

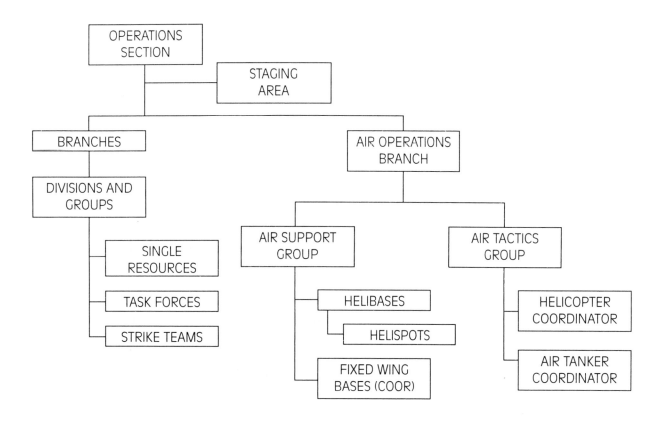

**Transparency #35** *Operators section.* (Taken from Figure 13-2 in the text)

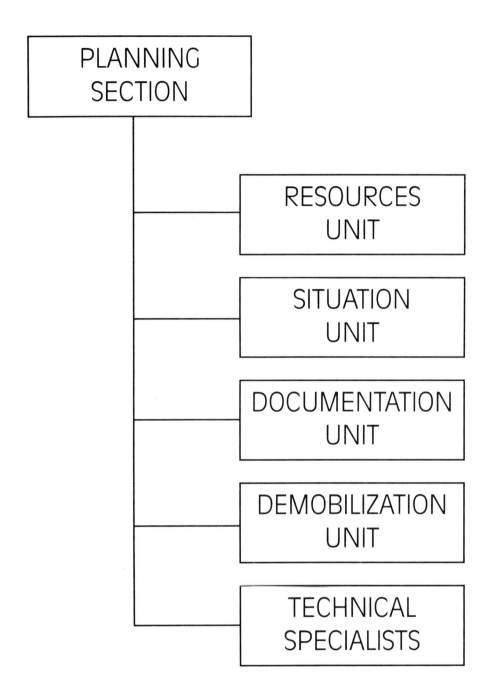

**Transparency #36** *Plans section.* (Taken from Figure 13-3 in the text)

Copyright © 1997 Delmar Publishers

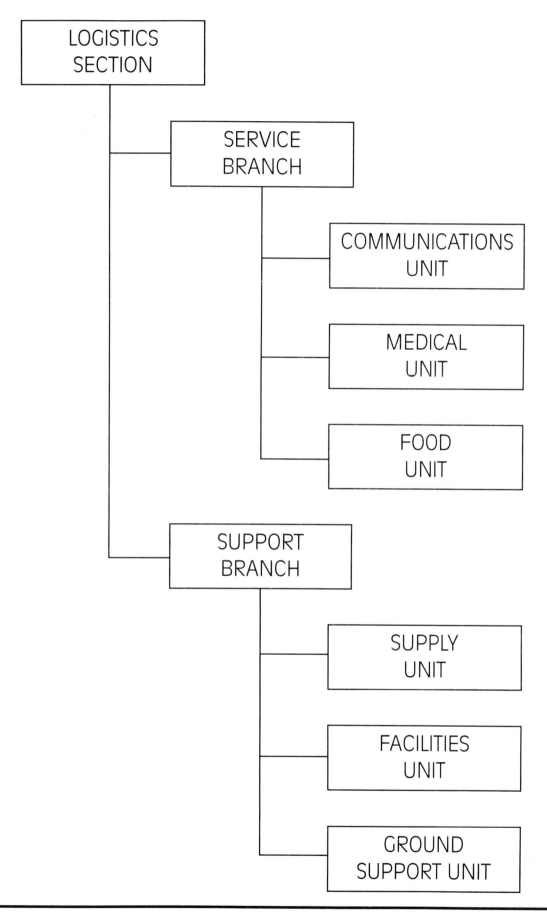

**Transparency #37**  *Logistics section.* (Taken from Figure 13-4 in the text)

Copyright © 1997 Delmar Publishers

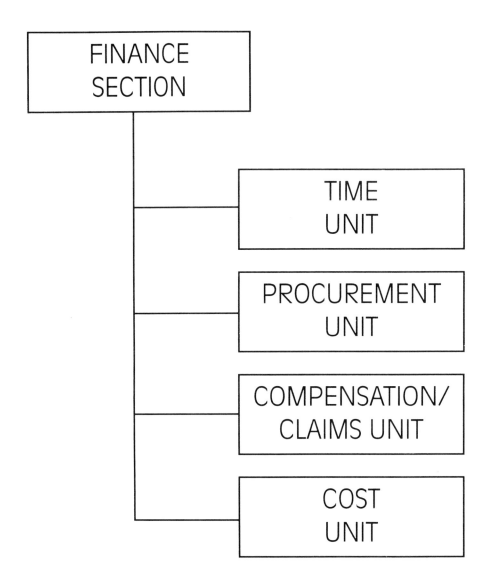

**Transparency #38** *Finance section.* (Taken from Figure 13-5 in the text)

## The Ten Standard Fire Fighting Orders

1. **F**ight fire aggressively, but provide for safety first.
2. **I**nitiate all action based on current and expected fire behavior.
3. **R**ecognize current weather conditions and obtain forecasts.
4. **E**nsure instructions are given and understood.
5. **O**btain current information on fire status.
6. **R**emain in communication with crew members, your supervisor, and adjoining forces.
7. **D**etermine safety zones and escape routes.
8. **E**stablish lookouts in potentially hazardous situations.
9. **R**etain control of yourself and your crew at all times.
10. **S**tay alert, keep calm, think clearly, and act decisively.

# The 18 Situations That Shout "Watch Out!"

1. The fire is not scouted and sized up.
2. You are in country not seen in daylight.
3. Safety zones and escape routes are not identified.
4. You are unfamiliar with the weather and local factors influencing fire behavior.
5. You are uninformed on strategy, tactics, and hazards.
6. Instructions and assignments are not clear.
7. No communications link with crew members/supervisor.
8. Building line without safe anchor point.
9. Building fire line downhill with fire below.
10. Attempting frontal assault on fire.
11. Unburned fuel between you and the fire.
12. Cannot see main fire, not in contact with anyone who can.
13. On a hillside where rolling material can ignite fuel below.
14. Weather is getting hotter and drier.
15. Wind increases and/or changes direction.
16. Getting frequent spot fires across the line.
17. Terrain and fuels make escape to safety zones difficult.
18. You feel like taking a nap near the fire line.

---

Copyright © 1997 Delmar Publishers

## TRIAGE

**T**ake time to evaluate water needs versus availability.

**R**econ safety zones and escape routes.

**I**s the structure defendable based on construction type, topography, and anticipated fire behavior?

**A**re flammable vegetation and debris cleared within a reasonable distance?

**G**ive a fair evaluation of the values at stake versus resources available and do not waste time on the losers.

**E**valuate the safety risk to the crew and the equipment.

# PROTECTION

**P**ark engines backed in so a rapid exit can be made if necessary.

**R**emember to maintain communication with your crew and adjoining resources.

**O**n occasions when you are overrun by fire, use apparatus or structures as a refuge.

**T**ank water should not get below 50 gallons in case it is needed for crew protection.

**E**ngines should keep headlights on, windows closed, and outside speakers turned up.

**C**oil a charged $1\frac{1}{2}$-inch hose line at the engine for protection of crew and equipment.

**T**ry not to lay hose longer than 150 feet from your engine.

**I**t is important to keep apparatus mobile for maximum effectiveness.

**O**nly use water as needed and refrain from wetting ahead of the fire.

**N**ever sacrifice crew safety to save property.

**L**OOKOUTS

**C**OMMUNICATIONS

**E**SCAPE ROUTES

**S**AFETY ZONES

\* \* \* \* \*

LOOK UP

LOOK DOWN

LOOK AROUND

# RULES FOR SAFE OPERATIONS AROUND HELICOPTERS

1. Approach and depart helicopters from the side or front in a crouching position, in view of the pilot.
2. Approach and depart the helicopter from the downhill side to avoid the main rotor.
3. Approach and depart the helicopter in the pilot's field of vision, do not go anywhere near the tail rotor.
4. Use a chin strap or secure your head gear (hard hat) when working under the main rotor.
5. Carry tools horizontally, beneath waist level to avoid contact with the main rotor.
6. Fasten your seat belt when you enter the helicopter and refasten it when you leave. A seat belt dangling out of the door can cause major damage to the thin aluminum skin of a helicopter.
7. Use the door latches as instructed. Use caution around plexiglass, antennas, and any moving parts.
8. When entering or exiting the helicopter, step on the skid. If you place your foot next to the skid and the weight of the ship changes, it may run over your foot.
9. Any time you ride in a helicopter in a wildland fire situation, you are required to wear full PPE.
10. Do not throw articles from the helicopter as they may end up in the main or tail rotors.

# Chapter 15

# Answers to Student Manual

## CHAPTER 1

### QUESTIONS

1. False
2. True
3. False
4. True
5. D
6. A
7. C
8. B

### EXERCISES

#### Exercise 1

Answers will be specific to the individual student.

#### Exercise 2

1. $2 \times 10^1 + 5$
2. $4 \times 10^3 + 5 \times 10^2 + 7 \times 10^1 + 2$
3. $1 \times 10^2 + 3 \times 10^1 + 6$
4. $9 \times 10^4 + 3 \times 10^3 + 2 \times 10^1 + 7$
5. 999
6. 126.7
7. 22,972
8. 185
9. 15
10. 1.26
11. 9,191
12. 5.78
13. 975
14. 47.28
15. 121,800
16. 5.628
17. 67
18. 22
19. 1100
20. 792.07
21. 24
22. 22
23. 12
24. $3^3$
25. $2^3 \times 5$
26. $31 \times 1$
27. $5 \times 3^2 \times 2^3$
28. $7 \times 5^2 \times 2$
29. $5^2 \times 2^5$
30. $\frac{3}{4}$
31. $\frac{7}{10}$
32. $\frac{162}{61}$ or $2\frac{40}{61}$
33. 27
34. 24
35. 74
36. $\frac{5}{6}$
37. $\frac{11}{15}$
38. $\frac{47}{60}$
39. $\frac{3}{33}$
40. $\frac{27}{20}$ or $1\frac{7}{20}$
41. 0.875
42. 0.52
43. 0.025
44. 88%
45. 52%
46. 3%
47. 7
48. 11
49. 9.84
50. 7 feet and 13 feet
51. The smaller number is 15. The larger number is 51.
52. 1,500 gallons
53. 18.75 p.s.i.
54. 4:00 A.M. Thursday
55. 3,960 square feet
56. 670.82 feet

#### Exercise 3

Answers will be specific to the individual student.

### ASSIGNMENTS

Answers will vary according to locale or jurisdiction.

165

## CHAPTER 2

### QUESTIONS

1. B
2. D
3. E
4. A
5. C
6. 1. Fire Prevention Specialist
   2. Fire Hazardous Materials Program Specialist
   3. Dispatcher
   4. Other positions as applicable
7. 1. Firefighter
   2. Fire Protection Systems Engineer
   3. Fire Protection Systems Maintenance Specialist
   4. Other positions as applicable
8. True
9. False
10. False
11. True

### EXERCISES

#### Exercise 1

Alarm Company Dispatcher, Non-Fire Suppression, Private

Emergency Communications Dispatcher, Non-Fire Suppression, Public

Company Officer, Fire Suppression, Public

Driver Operator, Fire Suppression, Public

Firefighters, Fire Suppression, Public

Battalion Chief, Fire Suppression, Public

Plant Safety Officer, Non-Fire Suppression, Private

Hazardous Material Team, Fire Suppression, Public

#### Exercise 2

Answers will be specific to the individual student.

### ASSIGNMENTS

Answers will vary according to jurisdiction.

## CHAPTER 3

### QUESTIONS

1. E
2. A
3. C
4. D
5. B
6. C
7. B
8. C
9. A
10. B
11. A
12. 1. Prevent fires from starting
    2. Prevent loss of life in case fire does start
    3. Confine fire to place of origin
    4. Extinguish fire once it does start
13. False
14. True
15. False
16. True

### EXERCISES

#### Exercise 1

Will be specific to individual student.

#### Exercise 2

1. Horse-pulled steamer
   Horse-pulled hose wagon
   Horse-pulled ladder wagon
2. 1. Slow mobilization of fire fighting forces.
      Adequate staffing levels with specific areas of protection assigned to specific fire companies.
   2. Dependence on horses to pull equipment.
      Engine-powered equipment.
   3. Inadequate equipment, including ladders and hose streams incapable of reaching past the third floor of six-story buildings and different thread types on couplings of responding companies.
      Adequate ladder and hose streams available for specific needs of the response area and standardization of coupling threads.
3. 1. Building material requirements, including the use of noncombustible materials to reduce fire potential.
   2. Required setbacks from property lines to reduce exposure potential to fires on adjacent property.
   3. Requirements for onsite fire extinguishing capabilities to extinguish small fires.
4. 1. Standardization of hose threads to allow maximum use of combined fire fighting forces would increase capabilities and decrease time to extinguishment.
   2. Standards for ladders and hose streams to meet necessary requirements of the fire protection area in which they are used.

### ASSIGNMENTS

Answers will vary according to jurisdiction.

# CHAPTER 4

## QUESTIONS

1. C
2. D
3. A
4. B
5. less
6. increases
7. flash point
8. ignition temperature
9. energy and oxidizer or heat and oxygen
10. chemical chain reaction
11. C
12. D
13. C
14. D
15. False
16. True
17. False
18. True

## EXERCISES

### Exercise 1

The tank car with the vapor density greater than 1 would present the greater danger. The vapor will sink and lie along the ground seeking low spots in which to pool and possible ignition sources.

### Exercise 2

1. 21%, 1800 to 2200
2. 21%, over 2200
3. below 15%, superheated

### Exercise 3

1. radiation
2. convection
3. convection, direct flame impingement

### Exercise 4

1. backdraft
2. smoldering phase
3. oxygen
4. The other two sides of the fire triangle were present. The contents of the room were superheated and the room was full of combustible fire gases.
5. free burning
6. 1. puffing smoke
   2. no visible flame production
   3. blackened windows

### Exercise 5

1. flashover
2. free burning
3. a confined unventilated area
4. sudden increase in temperature, ribbons or tongues of fire making runs across the ceiling and retreating
5. immediate exit

## ASSIGNMENTS

### Assignment 1

Kerosene: 100–162, 410, 0.7–5.0, 0.8, 4.5

Diesel: 125, 505, , less than 1, greater than 1,

Alcohol: 53, 852, 2.5–12, 0.8, 2.07

Propane: 156, 842, 2.37–9.5, 1.52, 1.56

1. propane
2. No, the specific gravity of gasoline is less than water. Gasoline will float on water and continue to burn.
3. gasoline, kerosene, diesel, alcohol, propane
4. B
5. A. 1, B. 2, C. 2

### Assignment 2

1. Oil forms circular shapes on the surface of the liquid. Oil immediately reforms circular shapes and rises to the surface of the water.
2. Bubbles begin to form. There is a fizzing sound. The baking powder dissolves without stirring, leaving a residue that sinks to the bottom of the water.

# CHAPTER 5

## QUESTIONS

1. E
2. D
3. A
4. C
5. B
6. True
7. True
8. True
9. A
10. B
11. C
12. C

## EXERCISES

### Exercise 1

*PRE-INCIDENT SUPPORT*

1. Emergency Management Institute; training course in large-scale disasters.
2. Municipal Fire Departments; mutual aid agreements in place.
3. Utility Companies and Volunteer Agencies participating in disaster and incident command training.

*INCIDENT SUPPORT*

1. Medical personnel for medical assistance.
2. Federal Emergency Management Agency; assist in management efforts at the incident, activate USAR task forces.
3. Utility Companies to shut down gas and electric lines to the area.

*POST INCIDENT SUPPORT*

1. Counselors trained in critical incident stress debriefings.
2. Utility Companies to restore utilities.
3. State Medical Examiner's Office and coroner; identification of victims.

### Exercise 2

*PRE-INCIDENT SUPPORT*

1. Highway Patrol; procedures for closing roadways.
2. Department of Transportation; Emergency Response Guidebook.

*INCIDENT SUPPORT*

1. Law Enforcement.
2. Towing Company with heavy wreckers.

## ASSIGNMENTS

1. Answers will be specific to local area.
2. Answers will be specific to local area.
3. 1075
4. 115
5. Extremely flammable; will form explosive mixtures with air.
6. Vapors may cause dizziness or asphyxiation without warning. Contact with gas or liquified gas may cause burns, severe injuries, and/or frost bite.
7. • Call emergency response telephone number on shipping papers. If it is available, refer to appropriate telephone number listed on the back cover of guide.
   • Isolate spill or leak area immediately for at least 50–100 meters for small spills, 800 meters (½ mile) for large spills.
   • Keep unauthorized personnel away.
   • Stay upwind.
   • Keep out of low places.

## CHAPTER 6

## QUESTIONS

1. True
2. True
3. False
4. True
5. B
6. A
7. B
8. A
9. B
10. A
11. C
12. A
13. B
14. D
15. C
16. A
17. less
18. incipient phase, free burning phase

## EXERCISES

### Exercise 1

1. 2½" NS double male
2. 2½" NS to 3" IP increaser/adapter
3. 1½" NS double female
4. 1½" NS to 1" IP reducer/adapter
5. 1½" NS to 2½" Siamese
6. 2½" to 1½" wye

### Exercise 2

1. Mcleod
2. Pulaski
3. Double Bit Axe; chopping
4. Single Bit Axe; chopping
5. Shovel

### Exercise 3

A. casing or volute
B. discharge
C. impeller
D. vane
E. eye
F. shaft

### Exercise 4

1. bulldozers to establish fire control lines and possibly 4-wheel drive patrols
2. Mcleod: scraping, removing heavy duff and forest litter.
   Pulaski: grubbing blade used for digging out roots, removing brush, loosening ground in preparation for scraping. Axe side used for removing lower limbs of trees, chopping heavier fuels.
   Shovel: scraping, throwing dirt to cool hot spots.
   Combi: used as a shovel, grub hoe, and chopping tool.
   Chain Saw: fell trees, buck up limbs and create fire breaks in brush.
3. fixed wing aircraft for retardant drops in open areas and rotary wing aircraft in less accessible canyon areas
4. a positive displacement pump because they have the ability to pump air and are self-priming

### Exercise 5

1. Full PPE, including full turnouts, boots, helmet, SCBA
2. 24-foot extension ladder, 10-foot attic ladder
3. chain saw, axe
4. pike pole, rubbish hook

## ASSIGNMENTS

### Assignments 1, 2, and 3

Answers will vary according to jurisdiction.

## CHAPTER 7

### QUESTIONS

1. False
2. False
3. True
4. False
5. C
6. B
7. B
8. D
9. C
10. A
11. C
12. B

### EXERCISES

#### Exercise 1

Deputy Chief, Deputy Chief, Deputy Chief / Battalion Chief / Company Officer / Driver Operator, Firefighter

#### Exercise 2

Through channels, which would begin with the appropriate Deputy Chiefs. They would notify the Battalion Chiefs, who would brief the company officers, who would then notify their driver operators and firefighters.

#### Exercise 3

1. C
2. The company officer on Y Shift will talk to the company officer of Z Shift, who is ultimately responsible for making sure his company follows standard shift change procedures.

#### Exercise 4

Company Officer, Battalion Chief

### ASSIGNMENTS

1. Answers will vary according to jurisdiction.
2. Answers will vary according to individual student.

## CHAPTER 8

### QUESTIONS

1. True
2. False
3. True
4. False
5. True
6. False
7. D
8. A
9. B
10. B
11. B
12. A
13. A
14. A
15. The reporting party's address
16. 1. budget information
    2. statistical information from emergencies
    3. payroll information
    4. training information and records

### EXERCISES

#### Exercise 1

*INCIDENT SUPPORT*

1. Meals; providing food for 874 people
2. Medical care; for minor or major injuries
3. Provide sleeping bags and shelter for off-shift personnel

*PRE-INCIDENT SUPPORT*

1. warehouse/central stores
2. repair garage
3. radio shop
4. fire business management

#### Exercise 2

1. 348-122-19
2. 348-122-02

#### Exercise 3

1. T27, R32
2. T27, R33

### ASSIGNMENTS

Answers will vary according to jurisdiction.

# CHAPTER 9

## QUESTIONS

1. True
2. False
3. False
4. True
5. C
6. B
7. D
8. B
9. safety
10. zero tolerance
11. B
12. A
13. A
14. B
15. B
16. A
17. A
18. 1. encourage speed
    2. encourage accuracy
19. When too much focus is placed on time, competition develops between crew members and companies, and safety is sacrificed for speed.

## EXERCISES

### Exercise 1

1. Safety Issues
   1. Appoint a safety officer to prevent unsafe acts.
   2. Do not use flammable liquids.
   3. Establish a building evacuation plan.
   4. Designate one person to control and ignite the fire in the presence of the safety officer.
2. Training Areas
   1. Training in the use of personal protective gear.
   2. Training in escape routes.

### Exercise 2

1. Technical Training
   1. Wildland fire behavior; knowing how the fire can be expected to progress.
   2. Wildland firefighter safety; knowing safety procedures.
2. Manipulative Training
   1. Fire shelter deployment.
3. Physical Training
   1. Physical strength and endurance training; maintaining optimum health.

## ASSIGNMENTS

1–5. Answers will vary according to individual student or jurisdiction.
6. Bring it to the attention of your company officer and the safety officer.

# CHAPTER 10

## QUESTIONS

1. False
2. True
3. False
4. False
5. B
6. C
7. B
8. C
9. Prevention Message
   1. bill boards
   2. school programs
   3. fire prevention week programs
   4. Media Day
10. Cause Determination
    1. We must know what is causing the fire in order to know how to target our prevention efforts.
    2. It can uncover problems with processes or equipment that can be corrected.

## EXERCISES

### Exercise 1—Fire Department Inspection Reports

1. Sundae Best (p.171)
2. Curly Q (p.172)
3. Vacant Building (p. 173)
4. River View Market (p.174)
5. The Drug Store (p.175)

### Exercise 2

Home Fire Inspection Report will vary according to individual student.

### Exercise 3—Fire Department Incident Reports

1. Incident 3075 (p. 176)
2. Incident 3097 (p. 177)

## ASSIGNMENTS

Answers will vary according to jurisdiction.

# FIRE DEPARTMENT
# INSPECTION REPORT

DATE: 1/30/98

BUSINESS NAME: Sundae Best

GENERAL USE: Ice Cream Shop

OCCUPANCY CLASS: B  2   PROPERTY CLASS: 5  1  3

LOCATION: 1530 Main Street, River View
Street Address                City

NUMBER OF BLDGS: 01   NO. OF STORIES: 01   BASEMENT (B) ____
                                            CELLAR   (C) ____

MANAGER: Maria Gonzales           PHONE: 555-8243
OWNER:   Maria Gonzales           PHONE: 555-8243

## DISPOSITION

1-CORRECTED    2-WILL CORRECT    3-VIOLATION NOTICE ISSUED    4-CITATION
                                  CALL BACK NECESSARY          ISSUED

| BUILDING | YES/NA | NO/DISP | FIRE PROTECTION EQUIP | YES/NA | NO/DISP |
|---|---|---|---|---|---|
| 1. EXIT DOORS | x/ | / | 8. FIRE EXTINGUISHERS | x/ | / |
| 2. EXIT SIGNS | x/ | / | 9. AUTOMATIC SYSTEMS | /x | / |
| 3. EXIT CORRIDORS | x/ | / | 10. WET STANDPIPES | /x | / |
| 4. AISLE SPACING | x/ | / | 11. DRY STANDPIPES | /x | / |
| 5. OCCUPANT LOAD SIGN | x/ | / | 12. ALARM SYSTEMS | /x | / |
| 6. VERTICAL OPENINGS | x/ | / | 13. FIRE ASSEMBLIES | /x | / |
| 7. WARNING SIGNS | /x | / | 14. FIRE WALLS | /x | / |

| COMMON HAZARDS | YES/NA | NO/DISP | SPECIAL HAZARDS | YES/NA | NO/DISP |
|---|---|---|---|---|---|
| 15. ELECTRICAL | x/ | / | 22. GREASE HOODS–DUCTS | /x | / |
| 16. FURNACE/BOILER RM | /x | / | 23. FLAMMABLE LIQUIDS | /x | / |
| 17. COOKING EQUIPMENT | /x | / | 24. LIQUID PROPANE GAS | /x | / |
| 18. HEATING EQUIPMENT | x/ | / | 25. COMPRESSED GAS | / | x/1 |
| 19. DECORATIONS | x/ | / | 26. CHEMICALS | /x | / |
| 20. HOUSEKEEPING | x/ | / | | | |
| 21. WEEDS | x/ | / | | | |

REMARKS: #25. Compressed gas cylinders for soda fountain need to be chained. Owner corrected in our presence.

CLEARANCE: GRANTED (G) _G_       INSPECTOR _____
           DENIED   (D) ____     OWNER/MANAGER _____

# FIRE DEPARTMENT
# INSPECTION REPORT

DATE: 1/30/98

BUSINESS NAME: Curly Q

GENERAL USE: Beauty Shop

OCCUPANCY CLASS: B 2    PROPERTY CLASS: 5 5 7

LOCATION: 1530 Main Street, River View
Street Address                City

NUMBER OF BLDGS: 01   NO. OF STORIES: 01   BASEMENT (B) ____
                                           CELLAR   (C)

MANAGER: Irene Johansen         PHONE: 555-2222

OWNER: Katie McClure            PHONE: 555-0124

## DISPOSITION

1-CORRECTED    2-WILL CORRECT    3-VIOLATION NOTICE ISSUED    4-CITATION
                                 CALL BACK NECESSARY           ISSUED

| BUILDING | YES/NA | NO/DISP |
|---|---|---|
| 1. EXIT DOORS | x/ | / |
| 2. EXIT SIGNS | x/ | / |
| 3. EXIT CORRIDORS | x/ | / |
| 4. AISLE SPACING | x/ | / |
| 5. OCCUPANT LOAD SIGN | x/ | / |
| 6. VERTICAL OPENINGS | / | x/2 |
| 7. WARNING SIGNS | /x | / |

| COMMON HAZARDS | YES/NA | NO/DISP |
|---|---|---|
| 15. ELECTRICAL | / | x/2 |
| 16. FURNACE/BOILER RM | /x | / |
| 17. COOKING EQUIPMENT | /x | / |
| 18. HEATING EQUIPMENT | x/ | / |
| 19. DECORATIONS | x/ | / |
| 20. HOUSEKEEPING | x/ | / |
| 21. WEEDS | / | x/2 |

| FIRE PROTECTION EQUIP | YES/NA | NO/DISP |
|---|---|---|
| 8. FIRE EXTINGUISHERS | / | x/2 |
| 9. AUTOMATIC SYSTEMS | /x | / |
| 10. WET STANDPIPES | /x | / |
| 11. DRY STANDPIPES | /x | / |
| 12. ALARM SYSTEMS | /x | / |
| 13. FIRE ASSEMBLIES | /x | / |
| 14. FIRE WALLS | /x | / |

| SPECIAL HAZARDS | YES/NA | NO/DISP |
|---|---|---|
| 22. GREASE HOODS–DUCTS | /x | / |
| 23. FLAMMABLE LIQUIDS | /x | / |
| 24. LIQUID PROPANE GAS | /x | / |
| 25. COMPRESSED GAS | /x | / |
| 26. CHEMICALS | /x | / |

REMARKS: #6. Hole in ceiling over the hair dryers needs to be repaired.

#8. Fire extinguisher must be mounted on wall.

#15. Extension cord to tanning booths must be replaced by permanent wiring.

#21. Weeds and trash outside back door by gas shut off need to be cleared.

CLEARANCE: GRANTED (G) _G_    INSPECTOR _____

           DENIED  (D) ____   OWNER/MANAGER _____

# FIRE DEPARTMENT INSPECTION REPORT

DATE: 1/30/98

BUSINESS NAME: Vacant Building

GENERAL USE: _____

OCCUPANCY CLASS: ___ ___     PROPERTY CLASS: ___ ___ ___

LOCATION: 1520 Main Street, River View
Street Address                City

NUMBER OF BLDGS: 01   NO. OF STORIES: 01   BASEMENT (B) ____
                                           CELLAR   (C) ____

MANAGER: _____   PHONE: _____

OWNER: _____   PHONE: _____

## DISPOSITION

1-CORRECTED    2-WILL CORRECT    3-VIOLATION NOTICE ISSUED    4-CITATION
                                 CALL BACK NECESSARY           ISSUED

| BUILDING | YES/NA | NO/DISP |
|---|---|---|
| 1. EXIT DOORS | / | / |
| 2. EXIT SIGNS | / | / |
| 3. EXIT CORRIDORS | / | / |
| 4. AISLE SPACING | / | / |
| 5. OCCUPANT LOAD SIGN | / | / |
| 6. VERTICAL OPENINGS | / | / |
| 7. WARNING SIGNS | / | / |

| COMMON HAZARDS | YES/NA | NO/DISP |
|---|---|---|
| 15. ELECTRICAL | / | / |
| 16. FURNACE/BOILER RM | / | / |
| 17. COOKING EQUIPMENT | / | / |
| 18. HEATING EQUIPMENT | / | / |
| 19. DECORATIONS | / | / |
| 20. HOUSEKEEPING | / | / |
| 21. WEEDS | / | / |

| FIRE PROTECTION EQUIP | YES/NA | NO/DISP |
|---|---|---|
| 8. FIRE EXTINGUISHERS | / | / |
| 9. AUTOMATIC SYSTEMS | / | / |
| 10. WET STANDPIPES | / | / |
| 11. DRY STANDPIPES | / | / |
| 12. ALARM SYSTEMS | / | / |
| 13. FIRE ASSEMBLIES | / | / |
| 14. FIRE WALLS | / | / |

| SPECIAL HAZARDS | YES/NA | NO/DISP |
|---|---|---|
| 22. GREASE HOODS–DUCTS | / | / |
| 23. FLAMMABLE LIQUIDS | / | / |
| 24. LIQUID PROPANE GAS | / | / |
| 25. COMPRESSED GAS | / | / |
| 26. CHEMICALS | / | / |

REMARKS: _____

CLEARANCE: GRANTED (G) _G_     INSPECTOR _____
           DENIED   (D) ____   OWNER/MANAGER _____

# FIRE DEPARTMENT INSPECTION REPORT

DATE: 1/30/98

BUSINESS NAME: River View Market

GENERAL USE: Retail Food

OCCUPANCY CLASS: B  2     PROPERTY CLASS: 5  1  1

LOCATION: 1500 Main Street, River View
Street Address           City

NUMBER OF BLDGS: 01   NO. OF STORIES: 01   BASEMENT (B) ____
                                           CELLAR   (C) ____

MANAGER: George Winn         PHONE: 555-4321

OWNER: Matt Johnson          PHONE: 555-1234

## DISPOSITION

1-CORRECTED     2-WILL CORRECT     3-VIOLATION NOTICE ISSUED     4-CITATION
                                   CALL BACK NECESSARY            ISSUED

### BUILDING

| | YES/NA | NO/DISP |
|---|---|---|
| 1. EXIT DOORS | / | x/2 |
| 2. EXIT SIGNS | x/ | / |
| 3. EXIT CORRIDORS | x/ | / |
| 4. AISLE SPACING | x/ | / |
| 5. OCCUPANT LOAD SIGN | x/ | / |
| 6. VERTICAL OPENINGS | x/ | / |
| 7. WARNING SIGNS | x/ | / |

### FIRE PROTECTION EQUIP

| | YES/NA | NO/DISP |
|---|---|---|
| 8. FIRE EXTINGUISHERS | / | x/2 |
| 9. AUTOMATIC SYSTEMS | / | x/1 |
| 10. WET STANDPIPES | /x | / |
| 11. DRY STANDPIPES | /x | / |
| 12. ALARM SYSTEMS | /x | / |
| 13. FIRE ASSEMBLIES | /x | / |
| 14. FIRE WALLS | /x | / |

### COMMON HAZARDS

| | YES/NA | NO/DISP |
|---|---|---|
| 15. ELECTRICAL | / | x/1 |
| 16. FURNACE/BOILER RM | /x | / |
| 17. COOKING EQUIPMENT | / | / |
| 18. HEATING EQUIPMENT | x/ | / |
| 19. DECORATIONS | x/ | / |
| 20. HOUSEKEEPING | / | x/2 |
| 21. WEEDS | /x | / |

### SPECIAL HAZARDS

| | YES/NA | NO/DISP |
|---|---|---|
| 22. GREASE HOODS–DUCTS | / | x/2 |
| 23. FLAMMABLE LIQUIDS | /x | / |
| 24. LIQUID PROPANE GAS | /x | / |
| 25. COMPRESSED GAS | x/ | / |
| 26. CHEMICALS | /x | / |

REMARKS: #1. Clear access to exit doors.
#8. Service due on fire extinguishers.
#9. Chain OS&Y valve in open position. Corrected in our presence.
#15. Clear storage in front of electrical panel. Corrected in our presence.
#20. Remove back storage to 18" below ceiling.

CLEARANCE: GRANTED (G) __G__     INSPECTOR _____

           DENIED   (D) ____     OWNER/MANAGER _____

# FIRE DEPARTMENT INSPECTION REPORT

DATE: 1/30/98

BUSINESS NAME: The Drug Store

GENERAL USE: Variety Store

OCCUPANCY CLASS: B 2    PROPERTY CLASS: 5 8 3

LOCATION: 1510 Main Street, River View
Street Address                City

NUMBER OF BLDGS: 01   NO. OF STORIES: 01   BASEMENT (B) ____
                                          CELLAR   (C) ____

MANAGER: Tom Smith                         PHONE: 555-2941

OWNER: Julie Michaels                      PHONE: 555-1246

## DISPOSITION

1-CORRECTED    2-WILL CORRECT    3-VIOLATION NOTICE ISSUED    4-CITATION
                                 CALL BACK NECESSARY          ISSUED

**BUILDING**

| | YES/NA | NO/DISP |
|---|---|---|
| 1. EXIT DOORS | x/ | / |
| 2. EXIT SIGNS | x/ | / |
| 3. EXIT CORRIDORS | x/ | / |
| 4. AISLE SPACING | x/ | / |
| 5. OCCUPANT LOAD SIGN | x/ | / |
| 6. VERTICAL OPENINGS | x/ | / |
| 7. WARNING SIGNS | /x | / |

**COMMON HAZARDS**

| | YES/NA | NO/DISP |
|---|---|---|
| 15. ELECTRICAL | / | x/2 |
| 16. FURNACE/BOILER RM | /x | / |
| 17. COOKING EQUIPMENT | /x | / |
| 18. HEATING EQUIPMENT | x/ | / |
| 19. DECORATIONS | x/ | / |
| 20. HOUSEKEEPING | / | x/1 |
| 21. WEEDS | /x | / |

**FIRE PROTECTION EQUIP**

| | YES/NA | NO/DISP |
|---|---|---|
| 8. FIRE EXTINGUISHERS | / | x/2 |
| 9. AUTOMATIC SYSTEMS | /x | / |
| 10. WET STANDPIPES | /x | / |
| 11. DRY STANDPIPES | /x | / |
| 12. ALARM SYSTEMS | /x | / |
| 13. FIRE ASSEMBLIES | /x | / |
| 14. FIRE WALLS | / | x/2 |

**SPECIAL HAZARDS**

| | YES/NA | NO/DISP |
|---|---|---|
| 22. GREASE HOODS–DUCTS | /x | / |
| 23. FLAMMABLE LIQUIDS | /x | / |
| 24. LIQUID PROPANE GAS | /x | / |
| 25. COMPRESSED GAS | x/ | / |
| 26. CHEMICALS | /x | / |

REMARKS: #8. Fire extinguishers are due for service.
#14. Repair hole in fire wall in bookkeeping area.
#15. Replace door panel on electrical box.
#20. Clear storage from in front of sprinkler valve.

CLEARANCE: GRANTED (G) __G__      INSPECTOR _____

           DENIED (D) ____         OWNER/MANAGER _____

# FIRE DEPARTMENT INCIDENT REPORT

GENERAL INFORMATION
YEAR/INCIDENT NUMBER  95-3075          INCIDENT DATE  1 0 1 0 9 5
DISPATCH TIME  1 3 2 8   END TIME  1 3 5 8    ALARM SOURCE  4
SITUATION FOUND  1 1    PROPERTY MANAGEMENT  1
INCIDENT ADDRESS/LOCATION  1259 Maple Street
CITY  River View                         GENERAL PROPERTY USE  4 1
SPECIFIC PROPERTY USE  4 1 1             OCCUPANCY TYPE  R . 3
STRUCTURED OCCUPIED AT TIME OF INCIDENT  1
OWNER NAME  Jason Smith
OWNER ADDRESS  1259 Maple Street
OCCUPANT NAME  Jason Smith

FOR MOBILE PROPERTY INVOLVED
TYPE  __   LICENSE NUMBER  _____    YEAR  __
MAKE  _____                 MODEL  _____

COMPLETE FOR ALL FIRES
ACTION TAKEN  1 3    FIRE ORIGIN AREA  4 3    LEVEL  A    FORM OF HEAT  1 5
IGNITION FACTOR  2 1   METHOD OF EXTINGUISHMENT  5
MATERIAL IGNITED FORM  6 5    TYPE  2 3    CONTRIBUTING FACTORS  3 1 6
PROPERTY LOSS  _____ 0     CONTENTS LOSS  4 5 0 . 0 0
ACRES BURNED  ____.0    FIRE CONTROLLED DATE  1 0 1 0 9 5    TIME  1 3 3 4

IF EQUIPMENT INVOLVED
TYPE  1 2    MODEL  A B C 2 C 8 3 3 T - L P G
SERIAL NUMBER  9 2 1 9 3 0 5 4 8 1

COMPLETE FOR ALL STRUCTURE FIRES
CONSTRUCTION TYPE  3    ROOF COVERING  2    NUMBER OF STORIES  0 1
EXTENT OF DAMAGE FLAME  2    SMOKE  2    SMOKE GENERATION TYPE  2 3    FORM  8 6

APPARATUS AND PERSONNEL

| UNIT RESPONSE | NUMBER PEOPLE | MILES ONE WAY | DISPATCH DATE/TIME | ARRIVAL DATE/TIME | RETURN DATE/TIME | RECOV TIME |
|---|---|---|---|---|---|---|
| E 1 | 0 0 3 | 0 0 1 | 1 0 1 0 9 5  1 3 2 8 | 1 0 1 0 9 5  1 3 3 1 | 1 0 1 0 9 5  1 3 5 8 | 3 0 |

COMMENTS:  Fire caused by improper storage of gasoline near a water heater. Discussed proper storage of flammable liquids with owner.

ACTIONS TAKEN  1 3

SIGNATURE: _____    DATE: _____

# FIRE DEPARTMENT INCIDENT REPORT

## GENERAL INFORMATION
YEAR/INCIDENT NUMBER  95-3097          INCIDENT DATE  101295

DISPATCH TIME  0155    END TIME  0315    ALARM SOURCE  4

SITUATION FOUND  14    PROPERTY MANAGEMENT  1

INCIDENT ADDRESS/LOCATION  Birch Road & Oak Street

CITY  River View                         GENERAL PROPERTY USE  65

SPECIFIC PROPERTY USE  655               OCCUPANCY TYPE  R.3

STRUCTURED OCCUPIED AT TIME OF INCIDENT  2

OWNER NAME  _____

OWNER ADDRESS  _____

OCCUPANT NAME  _____

## FOR MOBILE PROPERTY INVOLVED
TYPE  01     LICENSE NUMBER  CYN032     YEAR  94

MAKE  Chevrolet                MODEL  Pickup

## COMPLETE FOR ALL FIRES
ACTION TAKEN  15    FIRE ORIGIN AREA  81    LEVEL  A    FORM OF HEAT  64

IGNITION FACTOR  11    METHOD OF EXTINGUISHMENT  5

MATERIAL IGNITED FORM  86    TYPE  23    CONTRIBUTING FACTORS  263

PROPERTY LOSS  12000         CONTENTS LOSS  0

ACRES BURNED  .0    FIRE CONTROLLED DATE  101295    TIME  207

## IF EQUIPMENT INVOLVED
TYPE  __    MODEL  _____

SERIAL NUMBER  _____

## COMPLETE FOR ALL STRUCTURE FIRES
CONSTRUCTION TYPE __    ROOF COVERING __    NUMBER OF STORIES __

EXTENT OF DAMAGE FLAME __    SMOKE __    SMOKE GENERATION TYPE __    FORM __

## APPARATUS AND PERSONNEL

| UNIT RESPONSE | NUMBER PEOPLE | MILES ONE WAY | DISPATCH DATE/TIME | ARRIVAL DATE/TIME | RETURN DATE/TIME | RECOV TIME |
|---|---|---|---|---|---|---|
| E1 | 003 | 001 | 101295 0155 | 101295 0159 | 101295 0315 | 30 |

COMMENTS:  Vehicle fire; vehicle had been soaked with gasoline then ignited. A trailer of gasoline had been poured from the vehicle involved to another pickup which did not catch fire.

ACTIONS TAKEN  15

SIGNATURE: _____    DATE: _____

# CHAPTER 11

## QUESTIONS

1. False
2. True
3. True
4. True
5. False
6. False
7. True
8. True
9. True
10. False
11. True
12. True
13. E
14. C
15. A
16. B
17. B
18. C
19. C
20. B
21. A
22. C
23. D
24. A
25. B

## EXERCISES

### Exercise 1

Exterior attack

### Exercise 2

Ways to avoid another tragedy could be

1. The use of non-flammable materials for theater backdrops
2. Occupancy requirements consistent with exit capabilities
3. Exit doors that open outward instead of inward

## ASSIGNMENTS

Answers may vary according to jurisdiction.

# CHAPTER 12

## QUESTIONS

1. True
2. False
3. True
4. False
5. True
6. True
7. True
8. True
9. A
10. A
11. C
12. D
13. A
14. D
15. D
16. A
17. D
18. C
19. B
20. B
21. D & E
22. C
23. E
24. A
25. B
26. B
27. D
28. C
29. B
30. A
31. E
32. 
   1. flow rate
   2. inlet pressure
   3. length of line after the eductor
   4. elevation above the eductor
33. Its ability to dilute the oxygen content of the area, which can lead to asphyxiation.
34. cools and smothers
    insulates, cools, forms barrier
    smothers
35. aqueous film forming foam

## EXERCISES

### Exercise 1

1. upright head
2. pendant head

### Exercise 2

1. post indicator valve
2. OS&Y valve
   Purpose of valves is to make it obvious under quick inspection whether they are open or closed.

## ASSIGNMENTS

Answers will vary according to locale.

## CHAPTER 13

### QUESTIONS

1. True
2. False
3. False
4. True
5. False
6. True
7. False
8. False
9. A
10. C
11. B
12. C
13. A
14. B
15. E
16. C
17. A
18. D
19. rescue, exposures, confinement, extinguishment, overhaul, salvage, ventilation
20. facts, probabilities, your own situation, making a decision, planning the operation

### EXERCISES

#### Exercise 1 & 2

1. In the first incident one person, the battalion chief who was the highest ranking officer on scene, directed all fire operations, including assessment, attack, and safety issues. In the second incident the incident command system was used with different people assigned to various functions, including incident command, operations, and safety.
2. Allows the operation to be placed in a defensive mode instead of an offensive mode and uses available resources to protect exposures in an attempt to confine the fire to the area of origin.
3. Additional Areas
   1. Better communications, with separate channels for large incidents
   2. Accounting for personnel location and assignments

### ASSIGNMENTS

Answers will vary according to locale.

## CHAPTER 14

### QUESTIONS

1. B
2. D
3. C
4. B
5. C
6. C
7. False
8. True
9. False
10. False
11. True
12. False
13. False
14. False
15. True
16. B
17. A
18. C
19. upwind, uphill, upstream
20. isolate, identify, deny entry

### EXERCISES

#### Exercise 1

1. boil over
2. Water trapped below the level of oil heats up and expands to vapor at a rate of 1700 to 0, causing the oil to erupt outward.
3. defensive
4. evacuation of the area, use of remotely controlled master streams
5. foam, through; subsurface injection, or applied by aerial apparatus, or applied from the ground through large-bore nozzles that can project a stream over the wall of the tank.

#### Exercise 2

1. They usually happen on small fires or deceptively quiet sectors of large fires.
   They happen in light fuels, such as grass or brush.
   There is an unexpected shift in the wind direction or speed.
   They usually happen when fires run uphill.
   All four of these factors were present.
   Fire was described as an average creeping fire.
   Fuels were pinion juniper, oak brush, and short grass.
   Terrain was steep and rugged.
   A cold front moved in, changing wind speed and direction.
2. Safety zones and escape routes are not identified.

Unfamiliar with the weather and local factors influencing fire behavior.

Instructions and assignments are not clear.

On a hillside where rolling material can ignite fuel below.

Wind increases and/or changes direction.

Terrain and fuels make escape to safety zones difficult.

3. lookouts, communications, escape routes, safety zones
4. All relate to any incident to provide overall safe operations. Providing for safety and determining safety zones and escape routes for this creeping average fire that deceptively appeared nonthreatening, recognizing current weather conditions and obtaining forecasts would have all contributed to safer operations on the fire.

### Exercise 3

1. Exclusionary or "hot" zone
2. Contamination reduction or "warm" zone
3. "cold" zone

*PROTECTIVE EQUIPMENT*

1. Encapsulated suits, breathing apparatus, rubber gloves
2. Full turnouts, breathing apparatus
3. No special equipment

## ASSIGNMENTS

Answers will vary according to jurisdiction.